Exploring Mathematics
Investigations for Elementary School Teachers

(First Edition)

By
Rajee Amarasinghe
Lance Burger
Maria Nogin
Agnes Tuska
Oscar Vega

California State University, Fresno

cognella
San Diego, CA

Bassim Hamadeh, CEO and Publisher
Christopher Foster, General Vice President
Michael Simpson, Vice President of Acquisitions
Jessica Knott, Managing Editor
Kevin Fahey, Cognella Marketing Manager
Jess Busch, Senior Graphic Designer

First published in the United States of America in 2013 by Cognella, Inc.

Printed in the United States of America

ISBN: 978-1-62131-058-7 (pbk)/ 978-1-62131-059-4 (br)

✿ cognella™

www.cognella.com   800.200.3908

Contents

1 Problem Solving 1
   1.1 The Role of Problem Solving in Mathematics Teaching and Learning 1
   1.2 Polya's Four Steps of Problem Solving 2
       1.2.1 Polya's Four Steps Explained 2
       1.2.2 Polya's Four Steps in Action 4
       1.2.3 The Benefits of Alternative Solutions 5
   1.3 Investigations 7
   Investigation 1: Area of a Patio 7
   Investigation 2: Area with Geoboard 8
   Investigation 3: How Many Orders are There? 9
   1.4 Problem Solving 10
   Polya's Corner 10
   Exercises 10
   Chapter 1 Reflections 12

2 Fractions - Representation and Operations 13
   2.1 Teaching and Learning Fractions 13
   2.2 Visual Models for Fraction Representation 13
   2.3 Addition and Subtraction of Fractions 15
   2.4 Multiplication of Fractions 16
   2.5 Division of Fractions 19
   2.6 Investigations 21
   Investigation 4: What is $\frac{3}{4}$? 21
   Investigation 5: Equivalent Fractions Using the Number Line 22
   Investigation 6: Adding and Subtracting Fractions Using the Number Line 23
   Investigation 7: Pattern Blocks. Whole and Part 25
   Investigation 8: Addition and Subtraction with Pattern Blocks 26
   Investigation 9: Multiplication with Pattern Blocks 28
   Investigation 10: Division with Pattern Blocks 29
   2.7 Problem Solving 30
   Polya's Corner 30
   Exercises 30
   Chapter 2 Reflections 34

3 Number Sense 35
   3.1 What is Number Sense? 35
   3.2 Exponents, Number Bases and Place Value 35
   3.3 Investigations 40

Investigation 11: Memory or Understanding 40

Investigation 12: Sets of Numbers 41

Investigation 13: Prime and Composite Numbers 42

Investigation 14: Rational Numbers 43

Investigation 15: Representing the Real Numbers on the Real Number Line 44

3.4   Problem Solving 45

Polya's Corner 45

Exercises 45

Chapter 3 Reflections 48

4   Algebraic Thinking 49

4.1   What is Algebraic Thinking? 49

4.2   Inductive Reasoning and Algebraic Thinking 49

4.3   Investigations 51

Investigation 16: Algebraic Expressions 51

Investigation 17: Dot Patterns 52

Investigation 18: Square Patterns 53

Investigation 19: Staircase Problem 55

Investigation 20: Explorations With Patterns 56

Investigation 21: Number Patterns 57

Investigation 22: Area Model 59

Investigation 23: Making Sense of xy Algebra Tiles 60

Investigation 24: Addition of Algebraic Expressions Using xy Algebra Tiles 63

Investigation 25: Subtraction of Algebraic Expressions Using xy Algebra Tiles 64

Investigation 26: Multiplication of Algebraic Expressions Using xy Algebra Tiles 66

Investigation 27: Division of Algebraic Expressions Using xy Algebra Tiles 68

Investigation 28: Factoring of Algebraic Expressions Using xy Algebra Tiles 69

Investigation 29: Solving Linear Equations Using xy Algebra Tiles 70

Investigation 30: Solving Quadratic Equations Using xy Algebra Tiles 71

Investigation 31: A Balanced Diet? 73

Investigation 32: Comparing Solution Methods for Linear Equations 74

Investigation 33: Challenging Questions About Quadratic Equations 75

4.4   Problem Solving 76

Polya's Corner 76

Exercises 77

Chapter 4 Reflections 80

5   Geometric Thinking 81

5.1   Geometric Thinking and the van Hiele Levels 81

5.2   Understanding Geometry Through Measurement 82

5.3   The Pythagorean Theorem 84

5.4   Investigations 86

Investigation 34: Using Geoboards to Develop Perimeter and Area Concepts 86

Investigation 35: Exact Geoboard Perimeters 87

Investigation 36: Explorations with Pattern Blocks 88

Investigation 37: Circumference of a Circle 89

Investigation 38: Understanding the Pythagorean Theorem 90

Investigation 39: Interpretations of the Pythagorean Theorem 91

Investigation 40: Proofs of the Pythagorean Theorem 92

Investigation 41: Applications of the Pythagorean Theorem 94

Investigation 42: Developing Measurement Concepts 95

Investigation 43: Area of a Circle 97

Investigation 44: Volume 98

Investigation 45: Volumes of Power Solids 99

Investigation 46: Surface Area of Power Solids 100

Investigation 47: Surface Area 101

5.5 Problem Solving 102

Polya's Corner 102

Exercises 102

Chapter 5 Reflections 110

N   Exploring Resources 111

N.1 So, Why a Chapter N? 111

N.2 From Number Bases to Base-X Model. 111

N.3 Projects for Individual Explorations 113

N.4 Group Projects 115

Project 1: Numbers 115

Project 2: The Pythagorean Theorem 116

N.5 Multiple Representations 117

N.6 Rubric For Grading Problem Solving Tasks 118

N.7 Additional Challenges 119

N.8 Additional Challenges With Solutions 121

References 139

Preface

The authors of this book have collectively been teaching the *Exploring Mathematics* course for over ten years. It is an upper-division course intended as a culminating mathematics experience for liberal studies majors. After many changes to the course through much experimentation and collaboration, the authors have assembled a body of material meant to give the prospective teacher a highly conceptual understanding of mathematics topics essential for the elementary school level, by adopting the basic philosophy that the learning of mathematics takes time and is best learned from concrete models, multiple viewpoints and engaging problems. This text also focuses on the development of mathematical reasoning primarily through the use of manipulatives, models and visual aids, in addition to numerous investigations on topics prospective teachers typically have difficulty in teaching, such as fractions, algebraic thinking and geometry. The field-tested investigations and exercises found in this book will hopefully be as useful and enlightening for its students as it was for us, the authors, in the enjoyment of our own journey in learning and expanding mathematical pedagogy relevant to the educational challenges facing elementary school teachers today.

The Authors
Fresno, CA
April, 2012

Chapter 1
Problem Solving

1.1 THE ROLE OF PROBLEM SOLVING IN MATHEMATICS TEACHING AND LEARNING

> Solving problems is a practical art, like swimming, or skiing, or playing the piano: you can learn it only by imitation and practice... if you wish to learn swimming you have to go in the water, and if you wish to become a problem solver you have to solve problems.

— George Polya

Mathematics is about solving problems. Many students often believe they can't begin to solve problems because they're completely lost and don't know where to start. They tend to prefer to be shown a solution so that they could memorize it and apply the method to a similar problem hopefully occurring on a test. This might be a short-term remedy that helps for a *particular* problem, but doesn't contribute to gaining generalized problem solving techniques for a variety of situations.

Most states in the United States have now adopted the Common Core Standards for mathematics. The following eight Standards for Mathematical Practice not only guide the grade-level standards, but also promote a spectrum of good mathematical learning habits essential for problem solving proficiency.

The 8 Standards for Mathematical Practice

1. Make sense of problems and persevere in solving them.
2. Reason abstractly and quantitatively.
3. Construct viable arguments and critique the reasoning of others.
4. Model with mathematics.
5. Use appropriate tools strategically.
6. Attend to precision.
7. Look for and make use of structure.
8. Look for and express regularity in repeated reasoning.

This chapter focuses on content and investigations which encourage learners to find resourceful patterns in problem solving, as opposed to experiencing the general feelings of helplessness often associated with doing mathematics. Most experts don't immediately know how to solve math problems they encounter, but they have learned how to approach 'all' problems in similar ways. To begin learning how to do this; however, one must jump in 'head-first' into problem solving! If not, it's like trying to learn to ride a bicycle or learn to swim by being shown videos of other people doing so, but never practicing the actual activities yourself. Swimming looks so easy, but we know it isn't the first time and requires trial and error, instruction, understanding and practice. There is a large gap between 'watching' and 'doing', and learning mathematics is not very different. The next section begins the journey for prospective teachers to practice *systematic* approaches to problem solving, becoming examples which will one day help manifest the standards for mathematical practice in their own students.

1.2 POLYA'S FOUR STEPS OF PROBLEM SOLVING

Even though the tree of mathematics contains many different branches of vast topics, it so happens that there are common strategies one can learn for how to climb along those branches. But remember, in order to learn these strategies, you need to actually do the climbing yourself. Furthermore, for you as a prospective teacher, it is not enough to be able to solve problems yourself. Your task will include making sense of your students' attempts of solving problems and suggesting a variety of possible approaches to your diversely thinking students. This is why this book emphasizes the importance of multiple solution strategies encountered in problem solving, rather than giving numerous examples of procedures to be memorized. In his 1945 book *How To Solve It*, Polya identified the following four basic principles to guide problem solving:

Polya's Four Steps of Problem Solving

1. Understand the problem
2. Devise a plan
3. Carry out the plan
4. Look back

Before elaborating on the above steps, it is helpful to review Polya's 'Ten Commandments for Teachers' to better understand how to approach the teaching of problem solving:

1. Be interested in your subject.
2. Know your subject.
3. Try to read the faces of your students; try to see their expectations and difficulties; put yourself in their place.
4. Realize that the best way to learn anything is to discover it by yourself.
5. Give your students not only information, but also know-how, mental attitudes, the habit of methodical work.
6. Let them learn guessing.
7. Let them learn proving.
8. Look out for such features of the problem at hand as may be useful in solving the problems to come - try to disclose the general pattern that lies behind the present concrete situation.
9. Do not give away your whole secret at once - let the students guess before you tell it - let them find out by themselves as much as is feasible.
10. Suggest; do not force information down their throats.

1.2.1 Polya's Four Steps Explained

Step 1 - Understand the Problem

This first step can seem frustrating sometimes when first encountered by students, since if we 'understood' the problem, then everything would be easy, right? Well, Polya was a mathematician, and mathematicians are very precise in how they say things. The first step is to *understand the problem*, not understand how to solve the problem. So, it is not the case that if you understand the problem, you will necessarily know its solution. For example, if you want to paint a room and need to figure out how much paint to buy, there is a lot of understanding needed before one can get to the point of actually doing the computations that will find the answer you are looking for. How many coats of paint is one going to apply? How many square feet does one gallon of the paint cover in one coat? How many square feet of surface does one want to paint? What is the cost of the paint for each gallon? Understanding the problem begins with questions such as these. So, when beginning your understanding of a problem, try and begin forming questions such as:

- Do you know what the problem is asking?
- Do you know all of the terminology and symbols given in the problem statement?

- Do you know how the given information relates to what the problem is asking?
- Can you discuss and restate the problem to someone else?
- Is the problem like another one you might have already encountered?

Step 2 - Devise a Plan

After gaining a sense of what a problem is asking, the next step is to begin organizing problem information so that we can apply our thought and reflection. We do not want to be like the artist with a blank canvas, or the writer with the blank page, suffering from writer's block. We need to look at images, relationships, numbers, or any other relevant objects related to the problem so that we can begin to form a plan or strategy for how to solve the problem.

Some Strategies of Problem Solving When Devising a Plan

1. Guess and check.
2. Make lists.
3. Draw a picture or graph.
4. Look for a pattern.
5. Work backwards.
6. Make a table.
7. Assign variables to quantities.
8. Write an equation.
9. Find a formula.
10. Solve a simpler problem.

Step 3 - Carry out the Plan

At this stage, you have understood what the problem is asking and have assembled the information into some kind of mathematical form, which might be in terms of numbers, equations, graphs or pictures. After reflection and thought about the information given in the problem, a plan usually arises and begins forming. Once a plan is made, it is ready to be tested, or carried out. There is a saying in the game of Chess that "any plan is better than no plan at all." The same goes with solving mathematics problems. It is best not to judge your plan too harshly at first. If it does not work out, you may learn something from that, which can lead you to update or reject the plan for something better. One of the most common mistakes seen by teachers when their students attempt problem solving is that students often jump to the 'Carry Out the Plan' step before they have understood the problem and devised a plan. Doing this makes problem solving a lot harder since it will probably be much more difficult to evaluate the results of the plan if it was not based on a strategy developed from comprehension of what the problem was asking.

Step 4 - Look Back

After a plan has been put forward and implemented, it is time to reflect on the solution to determine if it has solved the problem. 'Looking Back' is easiest when the problem solver has thoroughly understood the problem, since the problem solver then has a good idea if the solution fits what the problem is asking. If the solution does not make sense or seem to solve the problem, it could mean that all previous problem solving steps need work. This points to a 'cycle' of problem solving in which the solver goes back to update their former understanding and strategies used to develop a plan, so that a new plan can be formed and checked against the updated understanding of the problem. Working in this way should ideally lead to a plan which, when checked, solves the problem. 'Looking Back' also includes looking for alternative, possibly simpler, strategies. In addition, you may 'play' with the given information and try to analyze under what conditions the problem would be solvable. This may lead to generalizations so that you will learn to solve a whole family of problems from the one problem you are working on.

1.2.2 Polya's Four Steps in Action

In order to demonstrate this four step problem solving process in action, let's consider the following problem:

There are ducks and rabbits in a yard. Together, they have 12 heads and 30 legs. How many of the animals are ducks and how many are rabbits?

Step 1 - Understand the Problem: It helps to visualize the problem situation. Of course, you have seen ducks and rabbits and know that they all have one head each, but ducks have two legs while rabbits have four. Notice that if all 12 were ducks, they would have only 24 legs, and if all 12 were rabbits, they would have 48 legs. So we know that there must be a mixture of the two kinds of animals in the yard.

Step 2 - Devise a Plan: There are different possible approaches for this problem. Here are a couple.

Plan A.

Let's organize all possible choices for the number of ducks and the consequences of those choices into a table.

number of ducks	number of rabbits	total number of legs
0	12	48
1	11	46
2	10	44
$\vdots$	$\vdots$	$\vdots$

Plan B.

Introduce a variable, say x, for the number of ducks. Express the number of rabbits in terms of x. Write the number of legs all the ducks and rabbits have, and then add them up. Since the total number of legs is given, we will obtain an equation. Then we will solve this equation for x, the number of ducks, and finally, figure out the number of rabbits.

Step 3 - Carry out the Plan:

Plan A.

number of ducks	number of rabbits	total number of legs
0	12	48
1	11	46
2	10	44
3	9	42
4	8	40
5	7	38
6	6	36
7	5	34
8	4	32
9	3	30

Plan B.

Let x be the number of ducks. Then the number of rabbits is $12 - x$. Since each duck has 2 legs, the total number of duck legs is $2x$. Since each rabbit has 4 legs, the total number of rabbit legs is $4(12 - x)$. Therefore the total number of all legs is $2x + 4(12 - x)$. So we have:

$$2x + 4(12 - x) = 30$$
$$2x + 48 - 4x = 30$$
$$18 = 2x$$
$$9 = x$$

Thus there are 9 ducks. It follows then that there are 3 rabbits.

Step 4 - Look Back:
Let's check our answer with a drawing.

This suggests another possible solution to the problem: draw 12 'heads' and 2 'legs' for each head. We will have only 24 legs then. But we need 30. So we must draw 6 more legs, thus 'transforming' 3 ducks into rabbits, leaving only 9 ducks.

If you wanted to write a similar problem, could you pick, say, 10 for the number of heads and 22 for the number of legs? How about 16 heads and 70 legs? How about 14 heads and 38 legs? What kind of numbers would work in general? What if you wanted to use different animals, say, ducks and octopuses? What can you say about the number of heads and the number of legs? Would 12 heads and 40 legs work?

Note that we could have chosen the variable x to represent the number of rabbits, rather than ducks. Then the number of ducks would be $12 - x$, and the total number of legs would be $4x + 2(12 - x)$. So we would obtain the following equation:

$$4x + 2(12 - x) = 30$$
$$4x + 24 - 2x = 30$$
$$2x = 6$$
$$x = 3$$

So there are 3 rabbits and therefore 9 ducks. Notice that this equation, and its solution ($x = 3$) is different from the one we had in Plan B above. This is because our variable x denotes a different quantity here; but, the answer to the problem is the same: 9 ducks, 3 rabbits.

1.2.3 The Benefits of Alternative Solutions

We all have strengths and weaknesses. Solving a problem in different ways can help us learn in our weak areas. For example, if you feel more comfortable working with tables than equations, solving the problem above in both ways may help you understand equations better. The answer should not depend on the method you used to solve a problem. If you solved a problem in two different ways and got two different answers, then you know that at least one solution is wrong.

Since our students often think differently, it helps if we can convey ideas in a variety of ways. It is also crucial for a teacher to learn how to understand various explanations that students come up with.

For example, 29×32 can be calculated in a variety of ways.

1. You may use the multiplication algorithm:

$$\begin{array}{r} 2\ 9 \\ \times\ 3\ 2 \\ \hline 5\ 8 \\ 8\ 7\ \ \\ \hline 9\ 2\ 8 \end{array}$$

2. You could think of 29 as $30 - 1$, so 29×32 is 32 less than 30×32:

$$29 \times 32 = (30 - 1) \times 32 = 30 \times 32 - 32 = 960 - 32 = 928$$

3. You could think of 32 as $2 \times 2 \times 2 \times 2 \times 2$, so

$$29 \times 32 = \underbrace{29 \times 2}_{58} \times 2 \times 2 \times 2 \times 2$$

$$58$$
$$116$$
$$232$$
$$464$$
$$928$$

4. You may visualize 29×32 as the area of a 29 unit by 32 unit rectangle, and relate it to the area of a 30 unit by 30 unit rectangle.

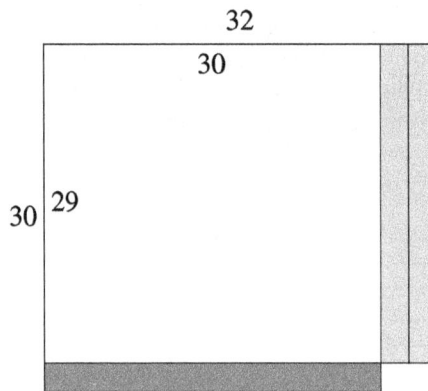

We need to add two of 1×29 rectangles, then take away one 1×30 rectangle. So the area of the 29×32 rectangle will be $2 \times 29 - 30 = 28$ more than the area of the 30×30 rectangle:

$$30 \times 30 + 28 = 928$$

1.3 INVESTIGATIONS

Investigation 1: Area of a Patio

The picture below shows Kim's patio (the distances shown are measured in feet).

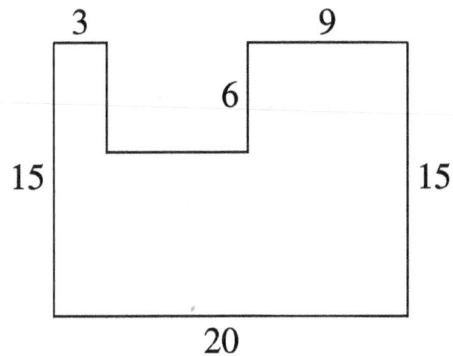

Kim wants to tile her patio with 1 ft × 1 ft ceramic tiles. She asked her friends to help her figure out how many tiles she will need. They sent her the following pictures.

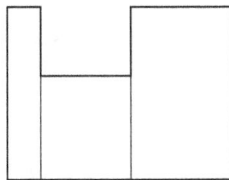
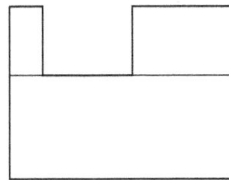
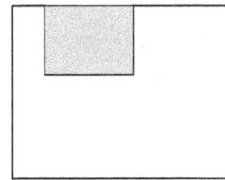

Plan A   Plan B   Plan C

Please help her to carry out each of the three plans.

Why was it not necessary to give the length of every segment in the figure? What assumptions did you have to make to solve this problem?

Investigation 2: Area with Geoboard

If we define the unit length as the distance between two closest points, then the area of the square in the picture below has area of $1 unit \times 1 unit = 1^2 unit^2 = 1 unit^2$ (we call it 1 square unit).

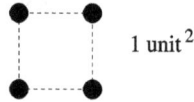

1 unit

$1 unit^2$

(a) Find the areas of the figures below enclosed by the (virtual) rubber bands. Do this in many different ways.

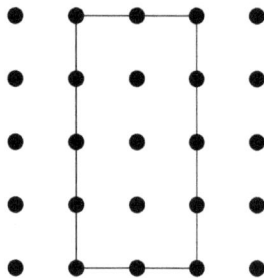

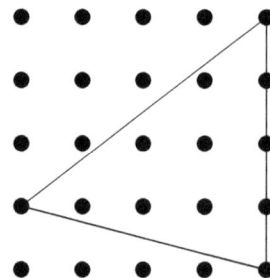

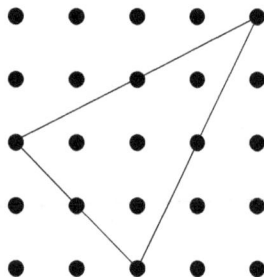

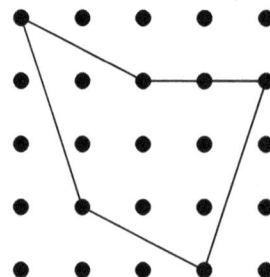

(b) Explain your methods for finding the area of each figure.

(c) After seeing several methods constructed by your classmates, do they all make sense to you? Which would be the easiest or fastest? Which would require perseverance?

(d) Create your own figures and then find their areas. Do this in many different ways. Explain your methods. Make sure you use correct mathematical vocabulary.

Investigation 3: How Many Orders are There?

Consider the following problem situation:

There are two kinds of large pizzas available in a pizza place. The veggie costs $12, the meaty $16. A group wants to spend exactly $136 on large pizzas because they are hungry, but that is all the money they have. What should they order?

(a) Write at least two different plans that might work to solve this problem.

(b) Try to find an order that would work by guess-and-check.

(c) Devise a plan that would allow you to find all of the possible orders that work.

(d) Solving a problem means to find its solution set. That set may be the empty set, may have one element, or may have many elements. How many elements does the solution set to this problem have?

(e) Construct a viable argument for the completeness of your solution set.

(f) Explain the difference between multiple ways of solving a problem and having multiple elements in the solution set of the problem.

1.4 PROBLEM SOLVING

Polya's Corner

The following *cookie jar problem* was used on many professional development workshops for teachers. Now it is your turn to try to solve it.

There was a jar of cookies on the table. Becky was hungry because she hadn't had breakfast, so she ate half the cookies. Then Jill came along and noticed the cookies. She thought they looked good, so she ate a third of what was left in the jar. Denise came by and decided to take a fourth of the remaining cookies with her to her next class. Then Jodi came dashing up and took a cookie to munch on. When Megan looked at the cookie jar, she saw that there were two cookies left. How many cookies were there in the jar to begin with?

1. Understand the problem. Tell the problem with your own words. What are you expected to do? What are your expectations about the number of cookies in the cookie jar?

2. Devise a plan. Try to find many different ways to answer the question.

3. Carry out the plan. Try to organize your work so that others could follow what you are doing.

4. Look back. Were your expectations about the number of cookies in the cookie jar met? What caused difficulty for you in this problem? What helped you in solving this problem?

EXERCISES

Solve problems 1-4 using Polya's 4 steps. Make sure to write down what you understand, what strategy you are planning to use, and explain the strategy. Try at least two different ways to solve each problem.

1. I am thinking about a number. One half is a third of my number. What is my number?

2. A balance scale was in perfect balance when Jane placed a box of candy on one pan of the balance and $\frac{3}{4}$ of the same-sized candy box together with $\frac{3}{4}$-pound weight on the other pan. How much did the full box of candy weigh?

3. Samuel was riding in the back seat of the station wagon on the way home after a long and tiring day at the beach. He fell asleep halfway home. He didn't wake up until he still had half as far to go as *he had already gone while asleep*. How much of the entire trip home was Samuel asleep?

4. Diana bought a piece of cloth 48 inches wide and 1 yard long. It cost $12. She cut off one-fourth and used it to make a tablecloth. From the remaining material, she used a piece that was 12 inches wide and 12 inches long to make a scarf and a piece 1.5 feet by 2 feet to make the cover for a pillow. Her sister saw Diana's sewing efforts and said she really liked the material. She wanted to buy what was left to do some sewing of her own. Diana was willing to sell the leftover material for the rate that she had paid for it. How much should she charge her sister?

5. An artist is planning to construct a rectangular wall design from square tiles. The wall is 72 inches long and 42 inches wide. All the square tiles must be the same size, and the length of the sides must be a whole number of tiles.
   (a) Find three different sizes of square tiles that could be used to completely fill the rectangular space, with no tiles overlapping and no tiles overhanging the border.
   (b) Determine the smallest number of square tiles that could be used to fill the rectangular space.

6. Find the area of the figure shown below. What strategies have you used?

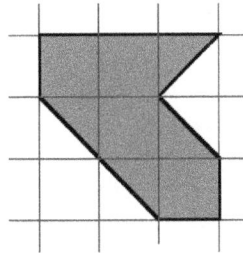

7. Find the area of the figures enclosed by rubber bands below. What strategies have you used?

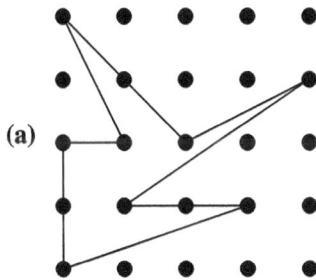

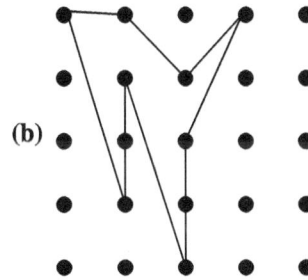

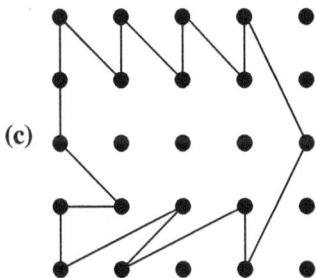

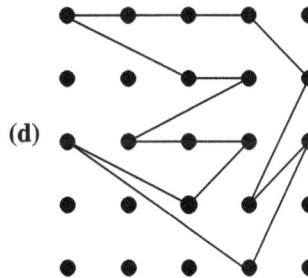

8. A group of nine people, some adults and some children, went to a movie theather. Adult admission was $8 and child admission was $5. The amount charged was $55. Could this amount be correct?

9. Give an example of a problem whose solution set contains
   (a) no elements
   (b) exactly one element
   (c) exactly four elements

CHAPTER 1 REFLECTIONS

1. Write about the role of problem solving in the mathematics classroom. Why should students practice problem solving? How can Polya's 4-steps help students to become better at problem solving? What are the various strategies students could use? How would solving the problem in multiple ways help students?

2. Explain the difference between multiple solutions of a problem and having multiple elements in the solution set of the problem.

3. How does the grading rubric you can find in chapter N help to promote problem solving in the mathematics classroom?

4. What is the role of manipulatives in the mathematics classroom? What are the major considerations when we use manipulatives?

5. Write about the usefulness and benefits, or possible difficulties and disadvantages, for the teacher and for the student of the discussion of errors.

6. Explain in your own words the meaning of each of the eight Standards for Mathematical Practice. Describe situations where you have encountered any of these practices in your study of Chapter 1.

Chapter 2
Fractions - Representation and Operations

2.1 TEACHING AND LEARNING FRACTIONS

In the previous chapter, we saw how the same problem may be solved by using multiple approaches. Similarly, using multiple representations for fractions can provide valuable ways to better understand and convey concepts such as: common denominator, addition/subtraction and multiplication/division of fractions. This chapter investigates models for representations and operations with fractions so that as teachers, you will be able to understand the arithmetic of fractions at a deeper level yourselves. Hence, you will be able to instill good learning habits of mathematics into the practices of your students. In this chapter we will focus on two Standards for Mathematical Practice, namely Modeling with mathematics (Standard 4) and Attending to precision (Standard 6), in the context of fractions. Remember, if you can understand a subject, you will naturally find the means to teach it - understanding is infectious and can produce knowledge, confidence and enthusiasm greatly contributing to student engagement.

2.2 VISUAL MODELS FOR FRACTION REPRESENTATION

The three most often used visual models to represent fractions are:

A. Area Diagrams
B. Number Line Diagrams
C. Set Diagrams

These diagrams help students to visualize fractions to better help them understand models for operations conceptually, instead of just formulas, like $\frac{a}{b} \cdot \frac{c}{d} = \frac{ac}{bd}$, to be memorized. As the Common Core Standards emphasize fraction concepts in grades 3–5, with the assumption of background knowledge of 'whole number' concepts developed in grades 1 & 2, number line and area diagrams are essential to convey fraction arithmetic based on their connection to the whole number concept.

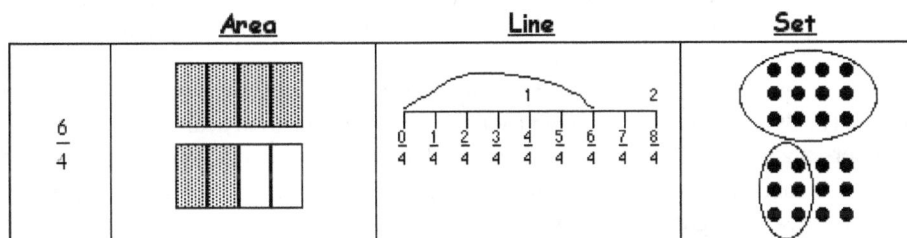

Area Diagrams

As seen in the figure above, one way to employ area diagrams for visual representation of fractional quantities involves creating rectangles by dividing the rectangle into the same number of parts as the denominator, while shading a numerator number of those subdivisions. For example, we see to represent $\frac{6}{4}$, two rectangles are subdivided equally into four pieces each, making $\frac{1}{4}$'s, and then 6 of those pieces are shaded.

It is important to realize that area diagrams for visual fraction representation does not insist upon only using rectangles. Any shapes can be used, as long as they are subdivided into equal-area parts. For example, you could use the following manipulatives to represent fractions using shapes other than rectangles.

- Pattern Blocks
- Pie Charts/Spinners
- Geoboards

It will be important to distinguish between area diagrams for visual representation of fractions and the area model for fraction operations, as in section 2.4, the area model for multiplication and division of fractions will be introduced, which *specifically* requires the use of rectangles.

Number Line Diagrams

In our view, a central reason the Standards for Mathematical Practice emphasize the use of number line diagrams, is that it can apply to a wide variety of common core topics, such as:

- Number ordering (greater than, less than, =, etc.)
- Rational numbers
- Decimals
- Approximate positions of irrational numbers (π, $\sqrt{2}$, etc.)

Method 1: Subdivided <u>Unit</u> Interval

Using the number line model to represent $\frac{3}{b}$ in a unit interval:

On the number line, the whole is the unit interval, which is the interval from 0 to 1, measured by length. To represent fractions such as $\frac{1}{2}, \frac{1}{3}, \cdots, \frac{1}{b}$, the whole is divided into b equal parts, so that each part has length $\frac{1}{b}$. We describe $\frac{3}{b}$ as represented in the number line as *three* of the $\frac{1}{b}$ lengths, as seen in the figure below

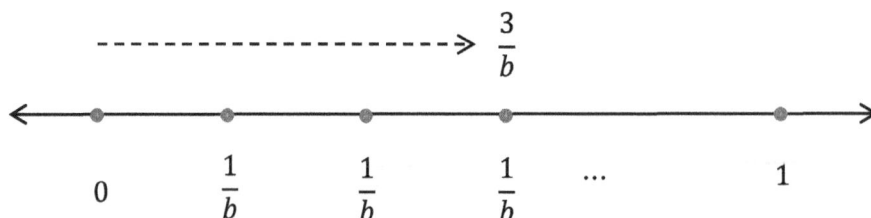

Method 2: <u>Any</u> Subdivided Interval, or 'Quotient' Method

The key idea for this variation of the number line model is to view $\frac{a}{b}$ not as the distance $\frac{1}{b}$ added up a-times; but, it interprets the representation $\frac{a}{b}$ as $a \div b$, which consists of the distance a on the number line, subdivided into b equal parts.

This might seem an insignificant difference, but it represents a different interpretation of the fraction since the previous method dealt with unit interval subdivisions; whereas, this method deals with the subdivisions over the entire numerator-length interval. In other words, to use the example of $\frac{3}{2}$; it is actually quite remarkable and significant that three $\frac{1}{2}$ subdivisions of the unit interval, is the same length as taking the interval of length 3, and dividing it into two pieces (see figure below). Why this works out to be equivalent, is the purpose of the models for fraction operations, which are closely connected to the models for fraction representations. Mathematical understanding is often the cumulative effect of many previous connected understandings, which can eventually contribute to an overall understanding; hence, it is recommended to use as many of the models and variants, such as methods 1 and 2, as possible in order to produce behaviors in students contributing to being flexible problem solvers.

$$3 \div 2 = \frac{3}{2}$$

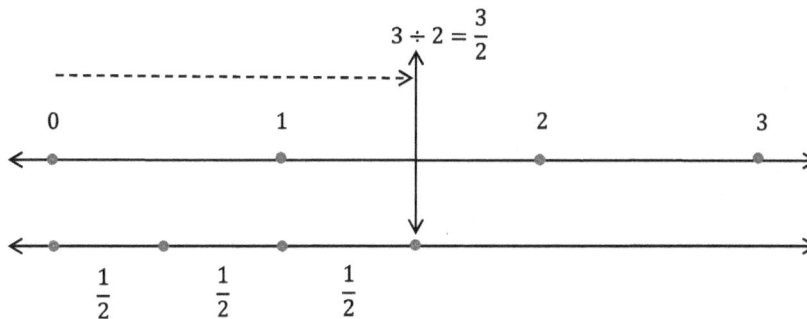

Another important use of number line diagrams is to allow students to easily visualize the *equivalence* of reduced fractions and their non-reduced counterparts, as seen in the following diagram depicting the equivalence of $\frac{1}{2} = \frac{2}{4} = \frac{4}{8}$.

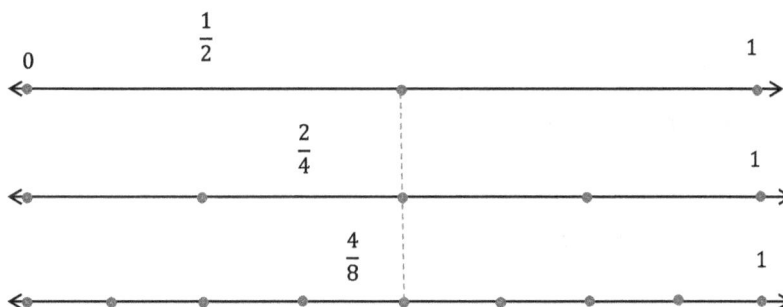

Set Diagrams

Set diagrams rely upon the use of individual objects which function as elements of a set making a 'whole', or 1. Since the entire collection of elements is a 'whole', then the fraction is the number of distinguished objects (numerator) 'of the whole' (denominator). In the above figure, as the circles are arranged into two groups of 12 pieces, then by segmenting the groups with ovals, the fraction $1\frac{6}{12}$ is represented, and can be concretely seen by the student that $1\frac{6}{12}$ must be equivalent to $1\frac{1}{2}$. Set diagrams can be used in any variety of ways using objects such as

- Colored Chips
- Toys
- Buttons
- Coins
- Integer Chips

Understanding fractions and their operations has typically shown to present challenging tasks for students. Mastering the operations without conceptual understanding can be one of the main reasons for these difficulties.

2.3 ADDITION AND SUBTRACTION OF FRACTIONS

In Common Core Standards-based instruction, students in grades 1 and 2 use the number line representation in order to make sense of addition and subtraction of whole numbers. The same representation, along with other visual representations, can be used to help learners construct viable arguments about the ways procedures should be generalized or modified for operations of fractions. For example, the need for a common denominator in adding or subtracting fractions arises in a natural way when number line, or pattern block, diagrams are used.

Based upon a variation of area diagrams for fraction representations, pattern blocks are geometric shapes which are subdivisions of the regular hexagon (a 6-sided polygon). For example, if we consider one hexagon to represent a whole (=1), then the values of pieces look like:

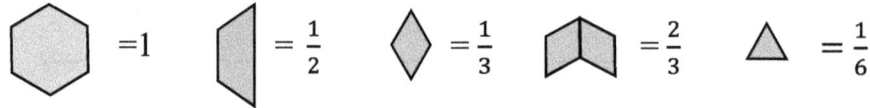

$$\text{⬡} = 1 \qquad \text{◗} = \frac{1}{2} \qquad \text{◇} = \frac{1}{3} \qquad \text{⬠} = \frac{2}{3} \qquad \triangle = \frac{1}{6}$$

But, if we now consider two hexagons to represent a whole (=1), then the pieces look like:

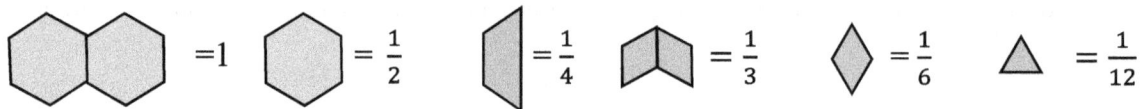

$$\text{⬡⬡} = 1 \qquad \text{⬡} = \frac{1}{2} \qquad \text{◗} = \frac{1}{4} \qquad \text{⬠} = \frac{1}{3} \qquad \text{◇} = \frac{1}{6} \qquad \triangle = \frac{1}{12}$$

In investigation 6 you will use number line diagrams, and in investigation 8 you will use pattern block diagrams for fraction addition and subtraction.

2.4 MULTIPLICATION OF FRACTIONS

Fraction Multiplication Using Rectangles

The concept of multiplication of fractions is directly connected to the multiplication of whole numbers. For example, just using whole numbers we know that 3×4 means 3 groups of 4 or 4 groups of 3.

$$3 \times 4 = 4 + 4 + 4 = 3 + 3 + 3 + 3$$

We can easily represent this idea using visual models:

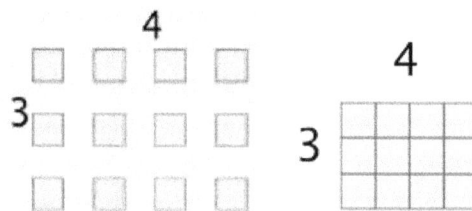

3 rows of 4 or as 4 columns of 3

This representation can be seen as a rectangle with side lengths of 3 and 4 (view this as a two number lines with lengths 3 and 4 as shown in the figure below). The answer for the multiplication is found as the area of the rectangle (number of units squares that cover the rectangle). In this case, the length of the rectangle is 3 units, the width is 4 units, and the area of the rectangle is 12 square units.

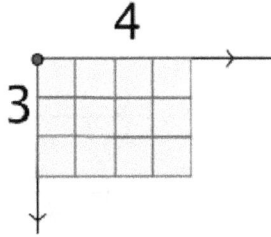

This idea can be used for any multiplication (whole numbers, polynomials, decimals or fractions), since multiplication is defined as repeated addition; therefore, when we can represent multiplication with a visual model using three simple rules (often called area model, box model, or array model).

Three rules for the visual area model for multiplication

1. Any multiplication can be represented as a rectangle
2. The two numbers/quantities that multiply represent the length and width of the rectangle.
3. The area of the rectangle represents the answer to the multiplication.

We can see in the following example that the same concepts that applied to whole number multiplication, can also apply to fraction multiplication. To illustrate the multiplication of any two fractions, use the above three rules to find the area of a rectangle, made of unit squares, so that fractional side lengths can be created.

Keep in mind that the area model for fraction operations is a little different than the area model for fraction representation, in that the model for representations is concerned solely with shading in a subdivided whole with correct fractional amounts; while, the area model for fraction operations is done by finding the area of a rectangle, given two fractions represented on number lines.

Example 2.1. $\frac{1}{2} \times \frac{1}{3}$

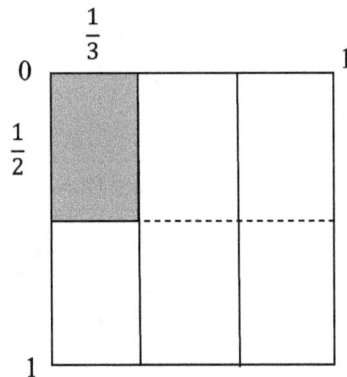

Notice here that the edges are the fraction factors that are multiplying and that the answer is the shaded *area*, which must be interpreted in terms of how many parts of 'the whole', meaning the 1×1 square. In this case, then, the answer is $\frac{1}{6}$ because we see that the whole has been divided into six equal parts and the rectangle of dimensions $\frac{1}{2} \times \frac{1}{3}$ has an area of $\frac{1}{6}$.

Example 2.2. $\frac{3}{4} \times \frac{3}{2}$

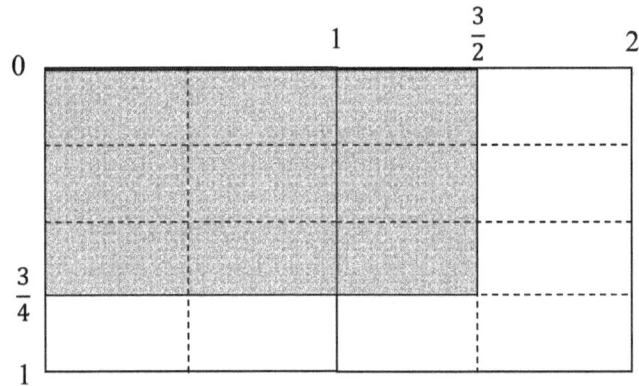

After finding the shaded rectangular region made by the $\frac{3}{4} \times \frac{3}{2}$ rectangle, we see that the area of the shaded region can be manipulated to fill in the two remaining squares on the bottom of the first 1×1 square and see that we have one small rectangle remaining, giving us the answer: $1\frac{1}{8}$ of a 1×1 square whole.

Fraction Multiplication Using Pattern Blocks

Using pattern blocks comprises another way to model fraction multiplication and is particularly useful for developing conceptual understanding for why the formula $\frac{a}{b} \times \frac{c}{d} = \frac{ac}{bd}$ works. The key to the pattern block approach for a problem such as $\frac{1}{2} \times \frac{1}{3}$ is to view $\frac{1}{2}$ groups of $\frac{1}{3}$ as *one of the two groups* of $\frac{1}{3}$. To do this with pattern blocks, though, it is necessary to first make *one of the two groups* of $\frac{1}{3}$ by dividing the $\frac{1}{3}$ pattern block representation into two equivalent groups of blocks. Finally, *one* of these two groups is taken which, in this case, ends up being $\frac{1}{6}$ relative to the given whole. This process is illustrated visually in the next example, and also in Investigation 9.

Example 2.3. Given that two hexagons is the whole ($= 1$), we want to use pattern blocks to diagram the multiplication

$$\frac{5}{2} \times \frac{3}{2}$$

Initial Drawing	Denominator Drawing	Numerator Drawing	Reduced Drawing	Answer
				$3\frac{3}{4}$

(a) The initial drawing shows $\frac{3}{2}$.
(b) The denominator drawing breaks $\frac{3}{2}$ in two parts due to the denominator of $\frac{5}{2}$ consisting of a two.
(c) Next, the numerator drawing depicts taking five (5) of those individual two pieces.
(d) For the final reduced drawing, one gathers up all of the 'wholes' and makes the remaining pieces all of one 'largest' color (the reduction).

2.5 DIVISION OF FRACTIONS

In the book, *Knowing and Teaching Elementary Mathematics* (1990), the mathematics education researcher, Liping Ma, shows how Chinese elementary school teachers develop their highly successful conceptual approaches to teaching. For the teaching of the topic of division with fractions, three basic models are employed:

1. measurement model
2. partitive model
3. factors and product (area) model

Measurement Model

The measurement model is based on the idea that if we want to perform a division, such as $6 \div 2$, it is the same as asking 'how many twos are in six?' One helpful way to convey the meaning of a model such as this is the use of a 'storyline' question which uses language and context to convey the model. For example, a storyline for $\frac{1}{2} \div \frac{1}{3}$ could consist of the questions:

- How many $\frac{1}{3}$ pounds of candy are in $\frac{1}{2}$ pounds of candy?
- A race is $\frac{1}{2}$ miles long. A lap is $\frac{1}{3}$ of a mile. How many laps do runners have to run in this race?

Partitive Model

The partitive model is based on the mathematical consequence that if $\frac{1}{2} \div \frac{1}{3} = x$, then it must be the case that $\frac{1}{3} \cdot x = \frac{1}{2}$. The following examples elaborate on the partitive model with storylines having similar content to the previous measurement model storylines:

- One-third of the weight of a box of candy weighs one-half of a pound. How much does the entire box of candy weigh?
- $\frac{1}{3}$ of the length of a race is $\frac{1}{2}$ of a mile. How long is the entire race?

Factors and Product (Area) Model

Another use of the area model is to compute a division, $\frac{A}{w} = \ell$, which consists in forming a 'leading edge', w, of the rectangle and extending the edge until the given area A has been accumulated, in which case the answer is the *top edge*, as in the following example for calculating $\frac{10}{2}$, we begin with a 'leading edge' of $2\,u$ and proceed to extend the edge out until we reach a total area of $10\,u^2$, in which case we see the *answer* in bold and underlined on the top edge.

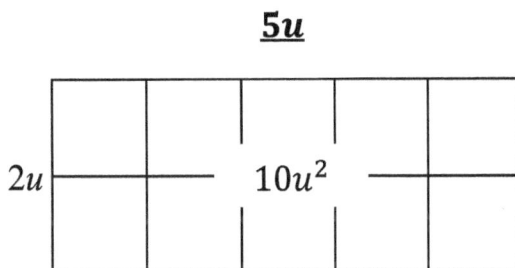

Similar to how we did division with numbers, we will now give an example of how to teach division with fractions using the area model:

Example 2.4. $\frac{1}{2} \div \frac{1}{3}$

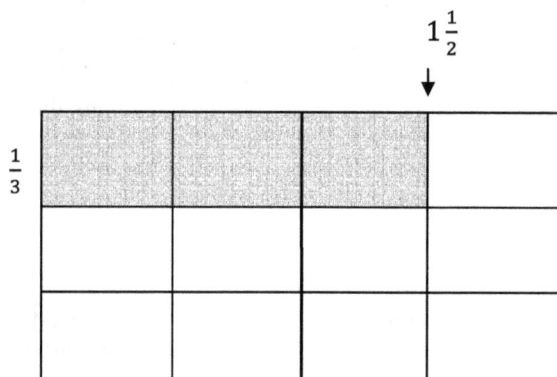

$1\frac{1}{2}$

$\downarrow$

$\frac{1}{3}$

To understand how the area model works for fraction division, the denominator $\frac{1}{3}$ is used as a leading edge for a rectangle that will be constructed to make an area of $\frac{1}{2}$. To do this in the above example, the side edges are subdivided into three parts. To accumulate an area of $\frac{1}{2}$ of a unit square, vertical half marks need to be made, which creates a common denominator of $\frac{1}{6}$ subdivisions of the unit square. When these common denominator subdivisions are made, it can be seen that three of the small $\frac{1}{3} \times \frac{1}{2}$ squares form half of the unit square. The area being constructed, though, must extend out horizontally, hence the three $\frac{1}{6}$ squares extend out $1\frac{1}{2}$ units of length relative to the 'top edge'. This points to one of the key elements of the model, to not confuse area with edge. The answer when computing with the area model is the 'top edge'.

2.6 INVESTIGATIONS

Investigation 4: What is $\frac{3}{4}$?

The following are representations of $\frac{3}{4}$ given by students. Can you explain the students' thinking?

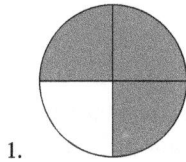

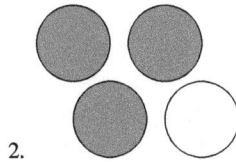

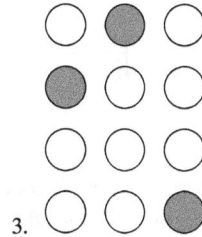

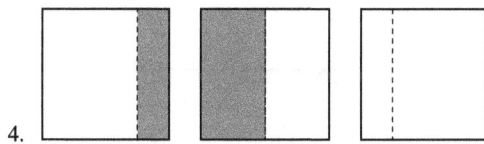

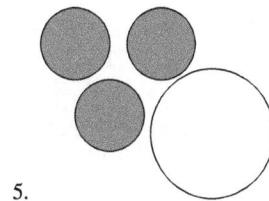

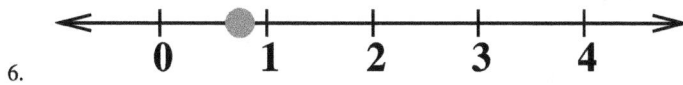

1.

2.

3.

4.

5.

6.

7. 0.75

8. I went to a store to buy 3 bottles of soda to divide equally among 4 people. How many will each get?

9. How many 4's are there in 3?

10. 18 out of 24.

Investigation 5: Equivalent Fractions Using the Number Line

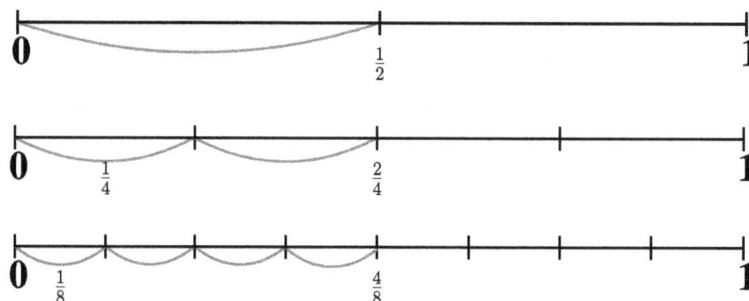

To do this process more accurately you can create 'Fraction Rulers' as used in common Japanese textbooks as follows. Figure 1 below shows how to represent any fraction with a denominator of a power of two (i.e 2, 4, 8, 16) by folding a strip of paper. Figure 2 shows how to find any fraction (not just a power of 2 in the denominator) by laying a strip across a ruled paper. This activity can help you to accurately compare fractions to determine which is bigger.

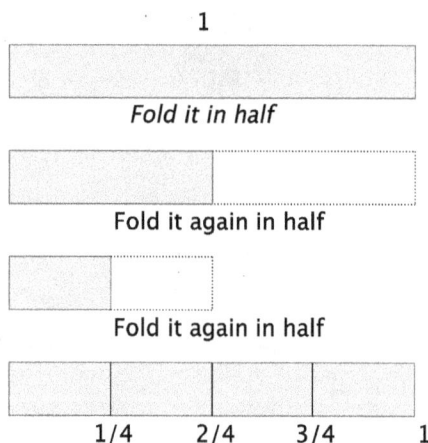

Figure 1

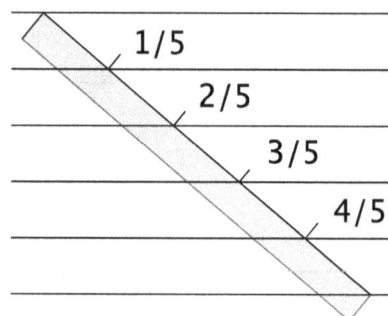

Figure 2

Using the number line, complete the following tasks.

(a) Show that $\frac{3}{4}$ is equivalent to $\frac{6}{8}$.

(b) Show that $1\frac{1}{2}$ is equivalent to $1\frac{4}{6}$.

(c) Show that $\frac{7}{3}$ is equivalent to $2\frac{1}{3}$.

(d) Compare the following fractions to determine which one is bigger: $1\frac{5}{6}$ and $1\frac{3}{4}$.

(e) Compare the following fractions to determine which one is bigger: $3\frac{1}{6}$ and $\frac{22}{7}$.

(f) Compare the following fractions to determine which one is bigger: $-\frac{72}{14}$ and $-\frac{41}{8}$.

Investigation 6: Adding and Subtracting Fractions Using the Number Line

1.

$$\frac{2}{7}+\frac{3}{7} = 2\times\frac{1}{7} + 3\times\frac{1}{7}$$

$$= \frac{1}{7}+\frac{1}{7} \ + \ \frac{1}{7}+\frac{1}{7}+\frac{1}{7}$$

$$= 5\times\frac{1}{7}$$

$$= \frac{5}{7}$$

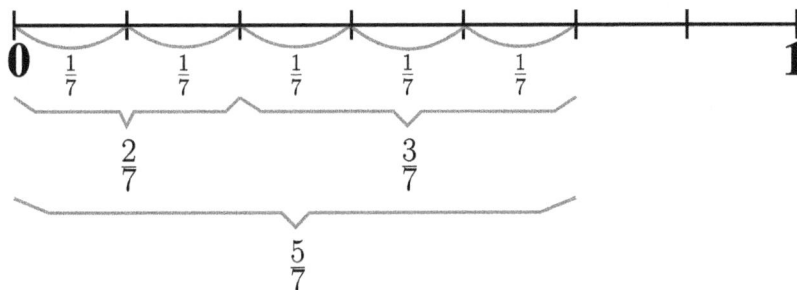

2.

$$\frac{3}{4}+\frac{2}{3} \ = \ \frac{1}{4}+\frac{1}{4}+\frac{1}{4} \ + \ \frac{1}{3}+\frac{1}{3}$$

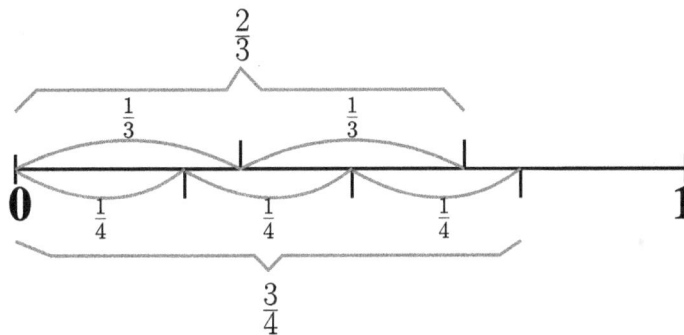

Finding common parts by dividing the whole into 12 equal parts, we get:

$$\frac{3}{4}=\frac{9}{12} \qquad\qquad \frac{2}{3}=\frac{8}{12}$$

$$\frac{17}{12} = 1\frac{5}{12}$$

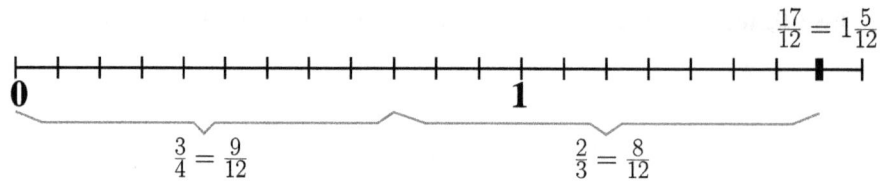

$$\frac{3}{4} = \frac{9}{12} \qquad\qquad \frac{2}{3} = \frac{8}{12}$$

$$\frac{3}{4} + \frac{2}{3} \;=\; \frac{9}{12} + \frac{8}{12} = \frac{17}{12} = 1\,\frac{5}{12}$$

3. Add the following fractions using the number line.

(a) $\dfrac{2}{5} + \dfrac{1}{5}$ (b) $\dfrac{2}{3} + \dfrac{3}{5}$ (c) $1\,\dfrac{2}{3} + \dfrac{3}{4}$ (d) $1\,\dfrac{3}{5} + 1\,\dfrac{2}{3}$

Investigation 7: Pattern Blocks. Whole and Part

1. If the trapezoid forms 'one whole,' then what fractions do the figures shown below represent?

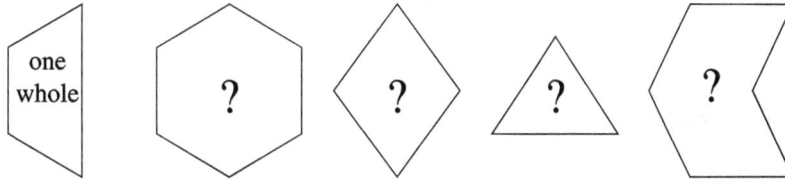

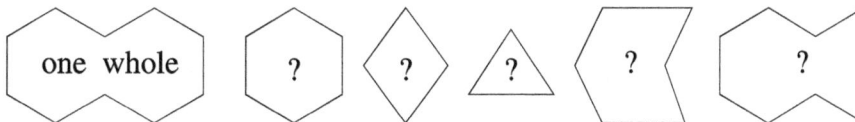

2. If two hexagons form 'one whole,' then what fractions do the figures shown below represent?

3. If the figures below indicate the fractions represented in their interior, find what is one whole.

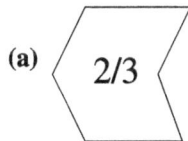

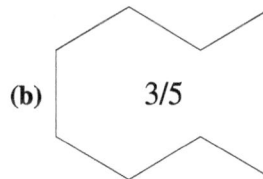

(a) 2/3   (b) 3/5

4. If two hexagons form 'one whole,' represent the following fractions using figures.

(a) $\frac{3}{4}$ (b) $\frac{5}{6}$ (c) $\frac{5}{12}$ (d) $\frac{2}{3}$ (e) $\frac{4}{3}$

Investigation 8: Addition and Subtraction with Pattern Blocks

Considering two hexagons as a 'one whole' how do we add the following two fractions?

$$\frac{2}{3} + \frac{3}{4}$$

Step 1: Represent the problem

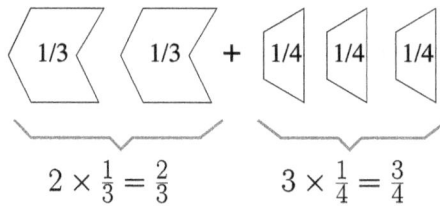

$$2 \times \frac{1}{3} = \frac{2}{3} \qquad 3 \times \frac{1}{4} = \frac{3}{4}$$

Step 2: Manipulation of fractions to add them. In order to add, all pieces have to be the same.

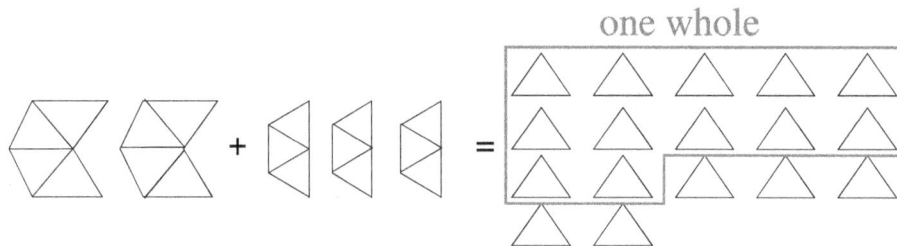

one whole

Step 3: Connect to addition algorithm.

$$\frac{2}{3} = \frac{8}{12} \qquad\qquad \frac{3}{4} = \frac{9}{12}$$

Finding the common denominator is similar to breaking the shapes into the same types of parts.

$$\frac{2}{3} + \frac{3}{4} = \frac{8}{12} + \frac{9}{12}$$
$$= \frac{17}{12}$$
$$= 1\frac{5}{12}$$

1. Consider two hexagons below as the whole. Using the above 3-step process, add the following fractions.

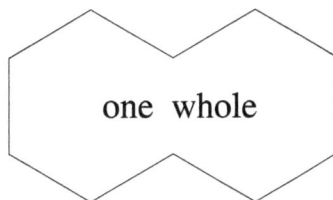

one whole

(a) $\dfrac{1}{2} + \dfrac{1}{3}$ (b) $\dfrac{2}{3} + \dfrac{1}{6}$ (c) $\dfrac{5}{6} + \dfrac{1}{2}$

2. Now consider the following figure as a whole and, using the above 3-step process, add the following fractions.

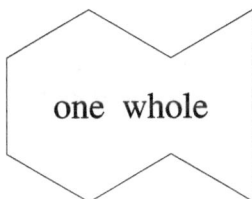

one whole

(a) $\dfrac{2}{9} + \dfrac{1}{3}$ (b) $\dfrac{2}{3} + \dfrac{5}{9}$

Investigation 9: Multiplication with Pattern Blocks

Consider two hexagons as one whole and find $\frac{2}{3}$ of $\frac{3}{4}$, and $\frac{3}{4}$ of $\frac{2}{3}$.

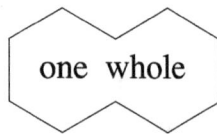

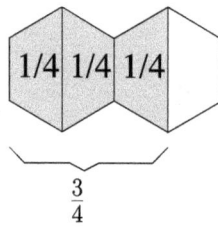

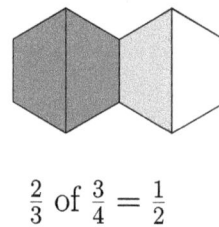

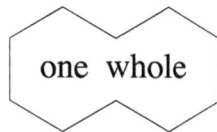

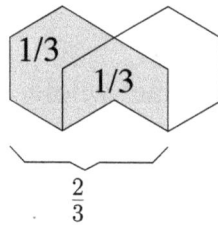

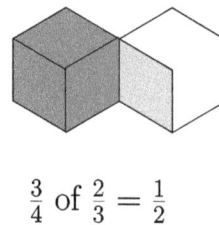

one whole   1/4 1/4 1/4

$\frac{3}{4}$

$\frac{2}{3}$ of $\frac{3}{4} = \frac{1}{2}$

one whole   1/3   1/3

$\frac{2}{3}$

$\frac{3}{4}$ of $\frac{2}{3} = \frac{1}{2}$

Using two hexagons as one whole find the following multiplications in two different ways.

1. (a) $\frac{1}{2} \times \frac{1}{3}$ (b) $\frac{1}{3} \times \frac{1}{2}$

2. (a) $\frac{2}{3} \times \frac{1}{4}$ (b) $\frac{1}{4} \times \frac{2}{3}$

3. (a) $\frac{5}{6} \times \frac{1}{2}$ (b) $\frac{1}{2} \times \frac{5}{6}$

Investigation 10: Division with Pattern Blocks

Considering two hexagons as a one whole represent the division of fractions $\frac{1}{2} \div \frac{1}{3}$.

$\frac{1}{2} \div \frac{1}{3}$ means the number of $\frac{1}{3}$'s there are in $\frac{1}{2}$.

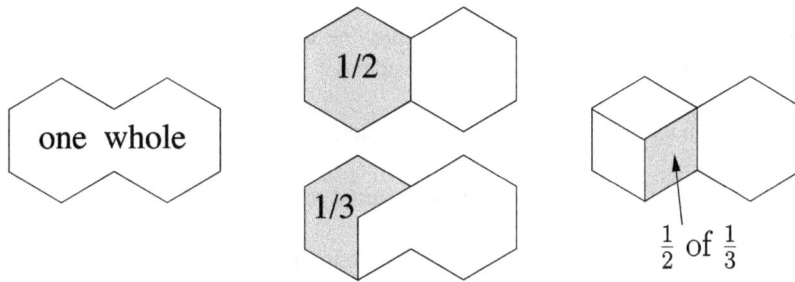

Therefore there are 1 and $\frac{1}{2}$ of $\frac{1}{3}$ in $\frac{1}{2}$.

With two hexagons as a whole draw pictures to find the answer to the following divisions.

(a) $2 \div \frac{1}{3} =$

(b) $\frac{1}{2} \div \frac{1}{4} =$

(c) $\frac{1}{3} \div \frac{1}{6} =$

(d) $\frac{3}{4} \div \frac{1}{4} =$

(e) $\frac{3}{2} \div \frac{3}{4} =$

(f) $\frac{5}{6} \div \frac{5}{12} =$

(g) $\frac{1}{2} \div \frac{1}{3} =$

(h) $\frac{3}{4} \div \frac{1}{2} =$

(i) $1 \div \frac{1}{2} =$

(j) $\frac{5}{6} \div \frac{1}{3} =$

(k) $\frac{2}{3} \div \frac{1}{2} =$

(l) $1\frac{1}{3} \div \frac{1}{2} =$

2.7 PROBLEM SOLVING

Polya's Corner

In a certain village $\frac{2}{3}$ of the men are married to $\frac{3}{5}$ of the women. Every one married male is married to one female and no one is married to anyone living outside the village. What portion of the population as a whole is married?

1. Understand the problem: Could a village, such as the one described in the problem, exist? If yes, would the village have more male or more female residents?

2. Devise a plan: Could you use a table or a diagram or a picture to visualize the situation in the village?

3. Carry out the plan: How could you write down the required ratio?

4. Look back: What can you say about the number of people living in the village?

EXERCISES

1. In each of the following parts, write a story problem involving the requested operation.
   (a) fraction addition.
   (b) fraction subtraction.
   (c) fraction multiplication.
   (d) fraction division.

2. Two-thirds of a fish weighs $10\frac{1}{2}$ pounds. How heavy is the whole fish?

3. A suit is on sale for $180. What was the original price of the suit if the discount was $\frac{1}{4}$ of the original price? Explain how you found your answer and how you can check your answer.

4. James uses $1\frac{1}{2}$ cups of milk and $2\frac{1}{4}$ cups of flour for his favorite cookie recipe. This makes 60 cookies. How much milk and flour would he need to make 40 cookies?

5. Write a story problem to describe each of the following

$$\textbf{(a)}\ 1\frac{2}{5} \times \frac{3}{4} \qquad\qquad\qquad \textbf{(b)}\ \frac{3}{5} \times 1\frac{2}{3}$$

Then choose an appropriate visual model to draw a diagram(s), and explain how to get the answer to your story problem.

6. Show the multiplication of the following numbers using the area model.
   (a) 26×314

   (b) $1\frac{2}{3} \times 2\frac{3}{5}$

   (c) For each computation above write a word problem (two distinct problems) for which you would have to do this calculation to get an answer.

7. What is $1\frac{3}{4} \div \frac{1}{2}$?

   (a) Show your calculation to get the answer.
   (b) Explain (using a picture or diagram or pattern blocks, etc.) to show what the problem and its solution represent.
   (c) Make up a word problem for which you would have to do this calculation to get an answer.

8. Write a story problem using the partitive model to describe

$$1\frac{1}{5} \div \frac{1}{4}$$

9. Write a story problem for division of fractions using the measurement model.

10. Write a story problem for division of fractions using partitive model.

11. Write a story problem for division of fractions using factors and product (area) model.

12. Write a story problem to describe each of the following

   (a) $\frac{1}{3} \div \frac{2}{5}$ (b) $\frac{3}{5} \div 1\frac{1}{2}$ (c) $\frac{3}{2} \div \frac{3}{4}$

   Then choose an appropriate visual model to draw a diagram(s), and explain how to get the answer to your story problem.

13. Use the area model to represent the multiplication of

$$1\frac{1}{3} \times \frac{3}{4}$$

   and find the answer using the diagram.

14. **Find at least two different solutions for the following problem.**
   The floor of a rectangular room is to be tiled with $\frac{1}{3}$ foot long, square-shaped tiles along a $9\frac{1}{4}$ foot wall. How many tiles will be needed along the wall?

15. **Find at least two different solutions for the following problem.**
   A land developer wants to develop 10 acres of land. Each lot in the development is to be $\frac{2}{9}$ of an acre. How many lots will the land developer have in the 10 acres?

16. Find three fractions such that they are between the following numbers. Then put the five numbers in increasing order. Explain your thinking.

   (a) $\frac{9}{35}$ and $\frac{1}{5}$ (b) $\frac{8}{21}$ and $\frac{1}{3}$ (c) $\frac{8}{35}$ and $\frac{1}{7}$ (d) $\frac{2}{5}$ and $\frac{3}{7}$

17. How many elements are there in the solution sets of the following problems?

- Problem 1: Find an integer that is more than $\frac{7}{4}$ but less than $\frac{9}{5}$.
- Problem 2: Find a fraction with denominator 100 that is more than $\frac{7}{4}$ but less than $\frac{9}{5}$.
- Problem 3: Find a fraction that is more than $\frac{7}{4}$ but less than $\frac{9}{5}$.

18. Consider the pattern blocks and the 'one whole' given below.

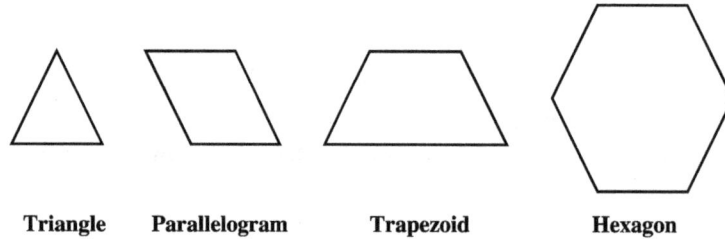

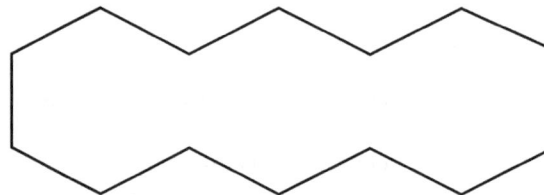

Triangle   **Parallelogram**   **Trapezoid**   **Hexagon**

One Whole

(a) What fractions of the whole above are the standard 4 pattern blocks? Write your answers inside the corresponding figures given above.

(b) Represent the following fractions using the whole given above, each in more than one way.

$$\frac{4}{9} =$$

$$\frac{1}{2} =$$

(c) Show how to find the sum $\frac{4}{9} + \frac{1}{2} = \frac{17}{18}$ by manipulating pattern blocks.

19. Consider the pattern blocks and the 'one whole' given in problem 18.

(a) Represent the following fractions using the whole given above, each in more than one way.

$$\frac{5}{6} =$$

$$\frac{1}{2} =$$

(b) Show how to find the sum $\frac{5}{6} - \frac{1}{2} = \frac{2}{6}$ by manipulating pattern blocks.

20. Consider the pattern blocks and the 'one whole' given in problem 18.
    (a) Represent the following two fractions using the whole given above, each in more than one way.

    $$\frac{4}{9} =$$

    $$\frac{1}{3} =$$

    (b) Show how to find the sum $\frac{4}{9} + \frac{1}{3} = \frac{7}{9}$ by manipulating pattern blocks.

21. Consider the pattern blocks and the 'one whole' given in problem 18.
    (a) Represent the following fractions using the whole given above, each in more than one way.

    $$\frac{5}{9} =$$

    $$\frac{3}{2} =$$

    (b) Show how to find the difference $\frac{5}{9} + \frac{3}{2} = 2\frac{1}{18}$ by manipulating pattern blocks.

22. Consider the pattern blocks and the 'one whole' given in problem 18. Use them to compute and explain the operations.

    (a) $\frac{1}{6} + 1\frac{3}{4}$ **(b)** $1\frac{5}{6} - \frac{1}{2}$

23. Try to re-do the previous 5 problems but now use the 'one whole' in the following picture. Discuss difficulties and benefits of using this new 'one whole'.

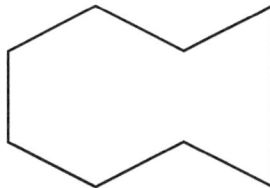

One Whole

24. Using two hexagons as one whole find the following multiplications in two different ways. Repeat this problem using four hexagons as one whole.

    (a) $\frac{1}{4} \times \frac{1}{3}$ and $\frac{1}{3} \times \frac{1}{4}$ **(b)** $\frac{3}{4} \times \frac{1}{3}$ and $\frac{1}{3} \times \frac{3}{4}$ **(c)** $\frac{3}{4} \times \frac{2}{3}$ and $\frac{2}{3} \times \frac{3}{4}$

25. Use the number line diagram and fraction rulers to decide which of the following two quantities is larger: three-fourths of 9 or two-thirds of 11.

26. We bought a pizza to share among friends. A third of it is just cheese, a half is pepperoni and the rest is half supreme and half veggie. If the whole pizza was $24, how much should the vegetarian pay for his slice?

27. Homer has 3 dozen donuts that he wants to share equally among 5 people. How much of a dozen can he give to each person?

    What is the 'whole' you used to solve this problem? Can you think of a different 'whole' that solves this problem?

28. President Obama has recently attended a ribbon cutting ceremony and as a souvenir he took $3\frac{3}{4}$ yards of ribbon. He wants to cut it into 3 equal pieces, one for his wife and one for each of his daughters. How long will each piece be?

29. **Find at least two different solutions for the following problem.**
    If 0.3 oz of mustard is used on each of 8 thousand hot dogs, how many 12-oz jars of mustard are needed?

30. Bob and Carla bought a pizza. Bob ate a third of the pizza and Carla ate a fourth of the pizza.

    a. How much pizza was left?
    b. If the weight of the remaining pizza is $\frac{5}{9}$ *lb*, what was the weight of the whole pizza?

CHAPTER 2 REFLECTIONS

1. Explain the reason why you need to find a common denominator when you add or subtract fractions.

2. Discuss the use of shortcuts in teaching mathematics (for example, using shortcuts for multi-digit number multiplication). Discuss what the implications in student learning are and what you can do as a future teacher.

3. Explain why any multiplication can be represented as a rectangle. Conjecture where you would find the two numbers you multiply and the product (answer) in that rectangle.

4. Explain why do you multiply by the reciprocal when you have to divide a fraction by another fraction.

5. Explain the reasoning of converting a mixed fraction into an improper fraction.

6. Write about your learning of fractions. It should include the following:

   • Summarize what you have learned about fractions in this section.
   • What new insights do you have about fractions?
   • What is still unclear about fractions?

Chapter 3
Number Sense

3.1 WHAT IS NUMBER SENSE?

Mathematics is in big part about numbers. Often, we use the term *number sense*, but what does this really mean? In the case of biology, number sense can refer to the ability of organisms to *sense* a change in the number of objects in a collection. In general, number sense is the ability to understand and recognize numbers, identify their relative values, and know how to use them in a variety of ways, such as in counting, measuring, or estimation. Number sense plays a significant role in the Grade 1 Common Core Standards, which emphasize the counting of numbers, as well as their relative magnitudes. In terms of the Common Core Standards of mathematical practice, number sense also plays important roles in the ability for students to *Attend to Precision* (Standard 6), as well as *Look for and Make use of Structure* (Standard 7). As numeracy and algebraic thinking have deep structural connections, this chapter explores number concepts such as number bases and place value, which ultimately build the foundation for algebraic thinking discussed in chapter 4.

3.2 EXPONENTS, NUMBER BASES AND PLACE VALUE

Writing a Check

To illustrate how structural features such as number and place value are already embedded in our daily lives and language, consider the case of writing a check. Let's suppose one wants to write a check for say, \$3,642.00. Most everyone who was to write this check would probably write something like, 'three-thousand six-hundred and forty-two dollars and no cents'. Let's examine this sentence in further detail by condensing the meaning of the words into mathematical notation as follows:

$$3 \cdot 1000 + 6 \cdot 100 + 4 \cdot 10 + 2 \cdot 1$$

Although the above expression is completely mathematical, we can condense it a bit more with the use of *exponents* representing repeated multiplication. More formally, we define for any number a:

$$a^n = \underbrace{a \cdot a \cdots a \cdot a}_{n \text{ times}}$$

So, given the use of exponents to represent repeated multiplication, we can condense the check-writing representation to:

$$3 \cdot 10^3 + 6 \cdot 10^2 + 4 \cdot 10^1 + 2 \cdot 10^0$$

Exponent Rule Justifications

One might be wondering why 10^0 is representing the number 1 in our previous check example. The reason for this is a consequence of the way that exponents are defined. As we have seen in chapter 2, the expression $\frac{a}{a}$, where $a \neq 0$, represents a number that, when multiplied by a, equals a. A fundamental property of our number systems is that there is only one number that does this, namely the number 1. In other words, 1 is the multiplicative identity for our number systems, i.e. $1 \cdot a = a$ for all numbers. This is just part of the reason that $a^0 = 1$. To look further into this, we need to investigate the properties of exponents.

Recall that when multiplying exponential expressions with the *same base* (10 in our check example), and for positive m and n,

$$a^n \cdot a^m = \left(\underbrace{a \cdot a \cdots a \cdot a}_{n \text{ times}} \right) \cdot \left(\underbrace{a \cdot a \cdots a \cdot a}_{m \text{ times}} \right) = \underbrace{a \cdot a \cdots a \cdot a}_{n+m \text{ times}} = a^{n+m}$$

Also, when we divide exponential expressions which share the same base we can see that: (assume that $0 \leq m \leq n$ and $a \neq 0$)

$$\frac{a^n}{a^m} = \frac{\overbrace{a \cdot a \cdots a \cdot a}^{n \text{ times}}}{\underbrace{a \cdot a \cdots a \cdot a}_{m \text{ times}}} = \underbrace{a \cdot a \cdots a \cdot a}_{n-m \text{ times}} = a^{n-m}$$

since m of the a's in the denominator cancel with the same number of a's in the numerator; hence, we want to subtract the m exponent from the n exponent. Thus, if $n = m$ and $a \neq 0$, we have $\frac{a^n}{a^n}$, which is the same number divided by the same number, and is thus equal to 1.

But, as a consequence of this exponent division rule, $1 = \frac{a^n}{a^n} = a^{n-n} = a^0$, which gives us the mathematical justification we were after for saying that $1 = 10^0$.

The other exponent division case, $m > n$, would be like an example such as:

$$\frac{a^2}{a^5} = \frac{a \cdot a \cdot 1}{a \cdot a \cdot a \cdot a \cdot a} = \frac{1}{a \cdot a \cdot a} = a^{2-5} = a^{-3}$$

Thus, to have this case be consistent with the rule $\frac{a^n}{a^m} = a^{n-m}$, it is a consequence that $a^{-n} = \frac{1}{a^n}$. Note; make sure you don't confuse these rules in cases like: $\frac{1}{a^2+a^3}$, which can only be simplified to an expression like, $\frac{1}{a^2(1+a)}$.

What about Dividing by Zero?

One of the most important things you can do for your student as mathematics teachers is to always try to provide justifications for the statements you present them, as it is of the utmost importance they see mathematical reasons why we say things like, 'you can't divide by zero'. Previously when justifying why $10^0 = 1$, it was mentioned that we could not let the base a be equal to 0. 0 is the additive identity for our number systems, namely, $0 + a = 0 + a = 0$ for all numbers a. In addition to being an additive identity, it is also the case that $0 \cdot a = a \cdot 0 = 0$ for all numbers a. To see this, let's reflect on the definition of multiplication as 'repeated addition'; for, just as exponents are a way to condense repeated multiplication, multiplication is a way to condense repeated addition.

$$n \cdot a = \underbrace{a + a \cdots a + a}_{n \text{ times}}$$

So, $1 \cdot 2 = 2$ is quite obvious from the above definition of multiplication since the 1 tells us to list just list a 2 one time. Isn't logical that if we say $0 \cdot 2$, this would imply listing the 2 'no times', which is equivalent to zero? And what about $2 \cdot 0$? Well, we could think of that as adding two of the 0's or $0 + 0$, but this is also equal to 0.

Dividing by Zero and Well-Defined Representations

In mathematics, it is very important for numerical expressions or representations, to be 'well-defined'. What this means is that the representation refers uniquely one and only one number. To illustrate this point with an example, on Tuesdays the symbol 3 still unconditionally means the same thing that it does on Saturdays, that is, the *cardinality* of a set containing three objects. This is not to say that we cannot have other representations for 3, like $\frac{6}{2}$. In fact, there are an infinite number of representations for 3; yet, each one of them *only represents* 3 and no other number. In this sense, they are called 'well-defined' representations. Then if $\frac{0}{0} = x$, and the expression $\frac{0}{0}$ is well defined, it is a unique number x such that $0 \cdot x = 0$. The only problem with this is that $0 \cdot x = 0$ for *every* number x. For this reason, we cannot get the expression $\frac{0}{0}$ to represent a unique number and so we therefore conclude that to divide by zero is *undefined* in order to maintain the soundness and consistency of our number systems.

Base-10

Now that we have more understanding of exponents, we could even use the base-10 bins to represent a dollar amount like $23,456.37 as:

$10^4 = 10000$	$10^3 = 1000$	$10^2 = 100$	$10^1 = 10$	$10^0 = 1$	$10^{-1} = 0.1$	$10^{-2} = 0.01$
2	3	4	5	6	3	7

Base-5

Since we are familiar with working in base-10, it can be helpful to consider another base, such as base-5, in order to put us in the position of better understanding common issues and new strategies when teaching children base-10. The nice thing about number bases is that they all share the same 'structure'. Number bases simply amount to grouping by different numbers. In base-10 we group with 10's, and when we have 1 group of 10's, we use *place value* to indicate how many of those 10 groups we have. For example, if one has 10 of the 10 groups of unit objects, then one has $10 \cdot 10 = 100 = 10^2$ of the objects. In the following example, a student of age 26 represents their age in base-5. Notice that there are no 125's in 26, so that digit is not necessary, but there is one 25 in 26 with 1 left over; hence, $26_{10} = 101_5$ (read as 'one-zero-one base 5'). What is your age in base-5?

$5^3 = 125$	$5^2 = 25$	$5^1 = 5$	$5^0 = 1$
	1	0	1

All of the familiar operations we have learned in base-10 translate to base-5, due to sharing the same number system structure. For example, to add $33_5 + 443_5$ we 'carry' to the next digit when reaching groupings of 5 in a previous digit. To compare base-5 to base-10, the number 4 in base-5 is like the number 9 in base-10.

$5^3 = 125$	$5^2 = 25$	$5^1 = 5$	$5^0 = 1$
	4	4	3
+		3	3
1	0	3	1

Similarly, to compute a subtraction like $123_5 - 34_5$ we 'borrow' groups of 5 from higher digits if we need to, just as we do when subtracting with the familiar base-10 algorithm

$5^2 = 25$	$5^1 = 5$	$5^0 = 1$
$\not{1}$ →0	$\not{2}$ → 1 + 5 = 6	3+5=8
−	3	4
	3	4

Using Polya's steps and 'looking back', we can verify that the above subtraction makes sense by performing the appropriate addition to check our answer (remember that every time one counts out 5 it is like reaching 10):

$5^2 = 25$	$5^1 = 5$	$5^0 = 1$
+1	3+1	4
+	3	4
1	2	3

You might be wondering why you see numbers such as 5, 6 and 8 appearing when doing the 'carrying' in the subtraction problem. This is a mere convention being used because we are so used to working in our own base-10. Since we have to carry, it will always be the case that 4 will be the highest possible number occurring when subtracting, but since we have not been raised learning our addition and subtraction tables in base-5, it is easier to work in our own base in this isolated case. If there was some planet in which base-5 was the natural base, people would learn their base-5 tables in elementary school and quickly say that $13_5 - 4_5 = 4_5$ without much thought, just as we can easily respond that $13 - 4 = 9$. Familiarize yourself and try to understand why the following base-5 addition and multiplication tables make sense. Maybe you will remember how difficult learning your addition and multiplication tables might have been when you were a young child!

+	0	1	2	3	4
0	0	1	2	3	4
1	1	2	3	4	10
2	2	3	4	10	11
3	3	4	10	11	12
4	4	10	11	12	13

×	0	1	2	3	4
0	0	0	0	0	0
1	0	1	2	3	4
2	0	2	4	11	13
3	0	3	11	14	22
4	0	4	13	22	31

Base-5 Multiplication

Now that we have some experience with base-5 arithmetic, our familiar algorithms for multiplication and division can be applied to base-5. Perhaps the key structural feature of our number systems which makes this possible is the *distributive property* for our number systems. To illustrate the distributive property we will employ an area model diagram:

$$3 \cdot (2 + 4) = 3 \cdot 2 + 3 \cdot 4 = 6 + 12 = 18 = 3 \cdot 6$$

×	2	4
3	$3 \cdot 2$	$3 \cdot 4$

An instructive way to introduce base-5 multiplication is by means of the area model, in which the sides are given in base-5 and the 'counted out' area is the answer:

Example 3.1. $12_5 \times 23_5$.

×	1	2	3	4	10	11	12	13	14	20	21	22	23
1	1	13	30					200					320
2	2	14	31		110					230			321
3	3	20	32				140					310	322
4	4	21	33	100					220				323
10	10	22	34			130					300		324
11	11	23	40				210						330
12	12	24	41		120					240			**331**

Notice, in the above example, that after counting for a while, a pattern was noticed and could be used to make the counting easier. Let's check the solution using the familiar algorithm:

5^3	5^2	5^1	$5^0 = 1$
		2^{+1}	3
	×	1	2
	1	0	1
+	2	3	
	3	**3**	**1**

Another multiplication example: $23_5 \times 44_5$ (try this on your own and justify the answer based on the familiar way that you multiply numbers, but in a context of base-5).

5^4	5^3	5^2	5^1	$5^0 = 1$
		2^{+2}_{+2}	3^{+2}_{+2}	3
	×		4	4
	2	0	4	2
2	0	4	2	
2	**3**	**0**	**1**	**2**

Base-5 Division

In the area model multiplication example above for $12_5 \times 23_5$, if we had wanted to model $331_5 \div 12_5$ we could have started with a 'leading edge' on the left hand side of the rectangle, and then count outwards with that side length until the amount of area accumulated is 331_5. If, when you are finished, you did not make a perfect rectangle, than either you counted wrong, or the teacher did not make the problem to have a zero remainder. If there was a remainder, this would be any excess squares hanging off the rectangle after 331_5 was reached. Similarly for division, we can apply our familiar approach to long division, but we must do all of our multiplications and subtractions in base-5.

```
              2   3
        ┌─────────────
  12₅   │ 3   3   1
    −    2   4
        ─────────
             4   1
        −    4   1
            ───────
                 0
```

One of the difficult aspects of doing base-5 long division relates to our unfamiliarity with the base-5 multiplication tables. In the above example, someone on a base-5 planet would know that $12_5 \times 2_5$ is the appropriate first multiplication to perform in order to get closest to 33_5 but for us we need to probably try one more multiple of 12_5 to be sure; $12_5 \times 3_5 = 24_5 + 12_5 = 41_5$... too much! In this chapter, we have seen how base-10 and base-5 have identical arithmetic structures, only differing in the way we group and make our digit representations for quantities. Place value is one of the most important elementary number concepts, and is often misunderstood by children, especially when a zero is involved as a very much needed place holder. Imagine if a zero is left out of $2,000,000$. What would a person rather have, $\$2,000,000$ or $\$200,000$?

3.3 INVESTIGATIONS

Investigation 11: Memory or Understanding

In this investigation we want to compare math understanding and just plain memory.
Work on parts 1 and 2 but do not look at part 3. When you are done with the first two parts work on part 3.

1. Try to memorize (on flash cards if you please) the 'math facts' below.

(a) $3 \times 2 = 1$	(b) $12 + 6 = 3$
(c) $5 \times 7 = 0$	(d) $8 + 4 = 2$
(e) $9 \times 4 = 1$	(f) $1 + 11 = 2$
(g) $6 \times 6 = 1$	(h) $4 + 12 = 1$
(i) $11 \times 8 = 3$	(j) $8 - 2 = 1$
(k) $2 \times 6 = 2$	(l) $11 - 7 = 4$
(m) $9 \times 8 = 2$	(n) $11 - 3 = 2$
(o) $7 + 2 = 4$	(p) $13 - 6 = 2$
(q) $3 + 7 = 0$	(r) $37 - 15 = 2$
(s) $10 + 3 = 3$	(t) $17 - 8 = 4$

2. Quiz yourself on how many of these 'math facts' you have memorized.

3. Note that all answers in the 'math facts' above are numbers from 0 to 4. Calculate the real answers for those operations and try to find a pattern. (Hint: think about remainders.) Now quiz yourself again on those 'math facts'.

Investigation 12: Sets of Numbers

1. Define the following types of numbers. Show examples for each type.

   (a) Natural numbers.

   (b) Whole numbers.

   (c) Integers.

   (d) Rational numbers.

   (e) Irrational numbers.

   (f) Real numbers.

   (g) Imaginary numbers.

   (h) Complex numbers.

   (i) Prime numbers.

   (j) Composite numbers.

2. Using a Venn diagram or tree diagram show how are these different types of numbers are related to each other. Use the diagram to answer whether the following statements are true or false.

   (a) Every integer is a whole number.

   (b) Every rational number is an integer.

   (c) All whole numbers are integers.

   (d) All odd numbers are prime numbers.

   (e) All prime numbers are odd numbers.

   (f) There is no number which is rational and irrational.

Investigation 13: Prime and Composite Numbers

Use your knowledge of prime numbers to answer the following questions.

1. Does a prime multiplied by a prime ever result in a prime? Does a nonprime multiplied by a nonprime ever result in a prime? Always? Sometimes? Never? Explain your answer.

2. Does a nonprime divided by a nonprime ever result in a nonprime? Always? Some times? Never?

3. Is it possible for an extremely large prime to be expressed as a large integer raised to a large power?

4. Show that it is impossible to have three consecutive natural numbers all of which are prime.

Investigation 14: Rational Numbers

Rational numbers (or irrational numbers) can be defined using two different definitions:
(1) Rational numbers are numbers that can be written as an integer divided by a non-zero integer.
(2) Rational numbers are numbers whose decimal representations are either terminating or infinite repeating.

1. How would you explain that these definitions are equivalent?

   (a) Can we convert any ratio of integers into a decimal number?

   (b) Can we convert any decimal number into a ratio of two integers?

2. Use the above discussion to answer the following questions. Which of the following is true? Explain your answer.

   (a) Every number that has a non terminating decimal expansion is irrational.

   (b) If the number M is a rational number, then 1/M must be a rational number also.

   (c) Any given real number is either rational or irrational.

3. For which of the questions in problem 2 was it sufficient to find an example or a counterexample?

4. Convert the following ratios to decimals. Try to do this without doing the long division.

$$\frac{3}{10}, \quad \frac{12}{100}, \quad \frac{3}{5}, \quad \frac{7}{20}, \quad \frac{12}{25}, \quad \frac{7}{8}, \quad \frac{13}{125}, \quad \frac{2}{3}, \quad \frac{1}{6}, \quad \frac{2}{7}, \quad \frac{5}{9}$$

   What type of ratios produce terminating decimals? Why?

5. Convert the following decimals to ratio of integers:

$$7.34, \quad -2.345, \quad 2.33..., \quad 3.4545..., \quad 12.34545..., \quad 3.35123123...$$

Investigation 15: Representing the Real Numbers on the Real Number Line

Any real number (rational or irrational) can be represented on the real number line.

1. Represent the following rational numbers on the real number line (try to do it in multiple ways and explain your methods):

$$\frac{1}{3}, \quad \frac{3}{2}, \quad 1\frac{2}{5}, \quad -2\frac{3}{4}$$

2. Represent the following decimal numbers on the real number line:

$$2.5, \quad -3.5, \quad 2.345, \quad -3.4$$

3. Draw the following two pairs of decimal numbers on the real number line and find a rational number and an irrational number between them:

(a) 3.47 and 3.46

(b) −3.256 and −3.2567

4. Estimate the location of each of the following irrational numbers on the real number line. Can you find the exact location?

$$\sqrt{2}, \quad \sqrt{3}, \quad \sqrt{5}, \quad \sqrt{10}, \quad \sqrt{13}$$

3.4 PROBLEM SOLVING

Polya's Corner
Carlos was to mail eight boxes, each with the same content, to eight different people. When he was done sealing all of the boxes, he realized that his address book was missing. Most probably he left it inside one of the boxes. Could he determine which box (if any) contains his address book by weighing the boxes on a balance scale? What is the least amount of weighing he has to do?

1. Understand the problem: Tell the problem with your own words.

2. Devise a plan: Can you succeed by making less than 5 weighing? Can you be even more efficient? What conclusion(s) can you make from a particular weighing?

3. Carry out the plan: Have you covered all possible outcomes of the weighing?

4. Look back: Are you sure you found the least amount of weighing? Can you justify your reasoning? How many weighing would be needed if Carlos had 9, 10, 11, or 12 boxes? Could you find a formula that gives the least number of weighing needed for any given number of boxes?

EXERCISES

Place value and base

1. Convert the following numbers to base 5

   (a) 12   (b) 32   (c) 40   (d) 628

2. Convert the following numbers from base 5 to base 10.

   (a) 12_5 (b) 32_5 (c) 40_5 (d) 323_5

3. Perform the following computations in base 5. Explain any 'mysterious', or different from standard operations, steps. Do NOT convert to base 10 to perform the computations, but you can do this to check your answers if you please.

   (a) $133_5 + 244_5$ (b) $32_5 \times 14_5$ (c) $222_5 + 134_5$ (d) $22_5 \times 33_5$

4. Perform the following computations in base 5. Explain any 'mysterious', or different from standard operations, steps. Do NOT convert to base 10 to perform the computations, but you can do this to check your answers if you please.

   (a) $233_5 - 144_5$ (b) $24_5 \times 23_5$ (c) $223_5 - 134_5$ (d) $30_5 \times 24_5$

5. **Martians I:** It is well-known that Martians have 7 fingers in each of their hands (no thumb!). Use this to make sense out of the following computations found in one of their ships
   (a) $58 + 39 = 93$
   (b) $3 \times 15 = 41$

6. **Martians II:** A different, more advanced, group of Martians also have 7 fingers, their computations seem different from the previous ones.. can you make sense out of them?
   (a) $32 + 40 = 102$
   (b) $3 \times 15 = 51$

7. **Martians III:** It is also well-known that Martians love candy. A sealed box with exactly 20 pieces of candy was found in a ship. Outside of the box it reads "24 count". How many fingers do these Martians have?

8. Use the base 10 manipulatives to calculate:

   (a) $254 + 348$ (b) $348 - 254$

9. Place the digits $1, 2, 3, 6, 7$, and 8 in the boxes below to obtain

$$\begin{array}{cccc} & \square & \square & \square \\ + & \square & \square & \square \\ \hline \end{array}$$

   (a) the greatest possible sum;
   (b) the smallest possible sum.

   Explain your strategy!

10. Place the digits $1, 2, 3, 6, 7, 8$ in the boxes shown below to obtain

$$\begin{array}{cccc} & \square & \square & \square \\ - & \square & \square & \square \\ \hline \end{array}$$

    (a) the greatest possible difference;
    (b) the smallest possible difference.

    Explain your strategy!

11. Assume that each letter represents a digit (different letters represent different digits). Find the value of each letter if you know that

$$\begin{array}{r} S\,E\,N\,D \\ +\,M\,O\,R\,E \\ \hline M\,O\,N\,E\,Y \end{array}$$

Building numbers

12. Using divisibility tests by 2, 3, 4, 5, and 9, explain how to determine whether a number is divisible by

    (a) 6   **(b)** 12   **(c)** 15   **(d)** 18

13. Which of 6, 12, 14, 15, 18 divide the following numbers? Provide an explanation for each answer. *Examples of explanations:* The number 4005 is not divisible by 6 because it is not divisible by 2. The number 4005 is divisible by 15 since it is divisible by both 3 and 5.

    (a) 3,210   **(b)** 3,528

14. Find prime factorizations of 3,210 and 3,528.

15. Find the greatest common factor and the least common multiple of 3,210 and 3,528.

16. The GCF of 66 and x is 11; the LCM of 66 and x is 858. Find x.

17. The GCF of m and n is 12, their LCM is 600, and both m and n are less than 500. Find m and n.

18. Given seven numbers $2, 3, 4, 5, 7, 10, 11$, pick five of them that when multiplied together give 2310. Find as many different solutions as you can.

19. A group of second grade students are playing the following game. They write digits from 1 to 9 in a row, and put a "+" or a "−" between every two consecutive digits. Then they calculate the result. For example,

$$1 + 2 + 3 - 4 + 5 + 6 - 7 + 8 - 9 = 5,$$

$$1 + 2 - 3 + 4 + 5 - 6 + 7 + 8 - 9 = 9$$

The goal is to come up with a sequence of $+/-$ signs for each answer between 1 and 10. (The person who first comes up with 10 sequences, one for each answer, will win.) Is it actually possible to do this?

20. Modify the game in the previous problem as follows: allow any order of the 9 digits, e.g.

$$4 + 8 - 1 + 5 - 7 + 3 + 6 - 2 - 9.$$

Will your answer to the question in the previous problem change?

21. What if in the game in problem 19 we allow 'combining' two or more consecutive digits, to form two- or more digit numbers, e.g.,

$$123 - 4 - 56 - 7 + 8 - 9$$

will now be allowed. Will your answer to the question in problem 19 change?

Real number line and decimals

22. Real numbers A, B, C, D, E, and F are represented by points on the number line below.

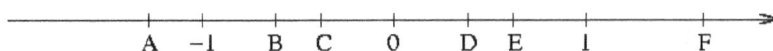

Determine the following if each answer is one of the numbers shown:

    (a) $D + E$ **(b)** $C + D$ **(c)** $B - D$ **(d)** E^2
    (e) BE **(f)** A^2 **(g)** $E \div D$ **(h)** $B \div F$

23. Daniel writes $0.4 < 0.13$ "because 4 is less than 13". Is he correct or wrong? Explain the correct reasoning
    (a) by drawing hundredths charts,
    (b) by a number line picture, and
    (c) by converting to fractions.
    (d) For which of your explanations does it help to write 0.4 as 0.40?

24. Find the 60th digit after the decimal point in the decimal representation of $\frac{5}{7}$. Explain your work using Polya's 4 steps of problem solving in your solution process.

25. What is the 100th digit after the decimal point in

    (a) $\frac{1}{7}$ **(b)** $\frac{1}{22}$ **(c)** 0.12112111211112...   **(d)** 0.1234567   **(e)** 0.1234567...

26. List 5 rational numbers between 23.83 and 23.84.

27. List 5 irrational numbers between 23.83 and 23.84.

CHAPTER 3 REFLECTIONS

1. Amy said "A rational number $\frac{a}{b}$ in its simplest form can be written as a terminating decimal if and only if the prime factorization of the denominator contains no prime numbers other than 2 and 5". Do you agree with her? Explain your reasoning.

2. Explain which kinds of numbers can be represented on the real number line. Discuss why it is useful for students to see how various numbers are represented on the number line.

3. Explain how learning bases other than base 10 can help you to prepare yourself to teach base 10.

Chapter 4
Algebraic Thinking

4.1 WHAT IS ALGEBRAIC THINKING?

The transitions from numeric to algebraic thought can present a variety of difficulties for many students. Often described as a 'cognitive gap', encountering the variable concept is a key source of roadblocks, which is one of the reasons the Common Core Standards aim to prepare students for algebraic thinking as early as Kindergarten. One way to do this is by developing inductive reasoning, such as in predicting patterns, earlier in the curriculum. Much like the cartoon in which a young student is depicted confused at the chalkboard looking at the answer $x = 2$, exclaiming to the teacher, "Now wait a minute, yesterday you said that x is equal to three!"; it is important for students not to see variables as just confusing notation which makes no sense. Consistent with the Common Core Standards for mathematical practices, this chapter promotes the use of manipulatives and visual models so students can better concretely see algebraic objects and operations in ways which allow them to make use of structure and repeated reasoning, important for making the leap in the generalization from arithmetic to algebraic thinking.

4.2 INDUCTIVE REASONING AND ALGEBRAIC THINKING

A key feature which distinguishes algebraic thought from numeric thinking is the ability for algebraic notation to condense and generalize mathematical information and patterns, as in the following example:

In the following pattern, creating number sentence expressions for each individual figure helps students to use inductive reasoning to develop generalizations using variables. You will also see inductive reasoning in Investigations 16 – 21.

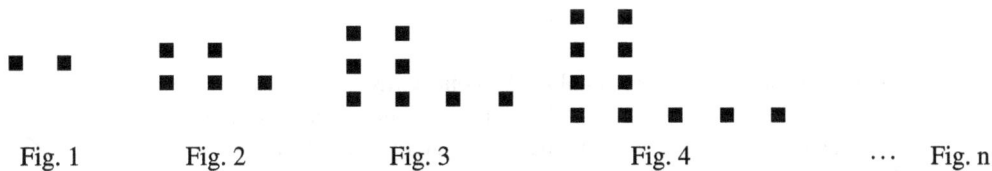

Fig. 1　　　Fig. 2　　　　Fig. 3　　　　　Fig. 4　　　　　　$\cdots$　Fig. n

Finding different ways to build number sentences describing patterns for how the number of dots changes from figure to figure leads to the development of algebraic thinking which can characterize a given pattern with variables. For example, we may use the following three strategies to obtain number sentences for each figure above

	Fig. 1	Fig. 2	Fig. 3	Fig. 4	$\cdots$	Fig. 10	$\cdots$	Fig. n
•	$2 \cdot 1 + 0$	$2 \cdot 2 + 1$	$2 \cdot 3 + 2$	$2 \cdot 4 + 3$	$\cdots$	$2 \cdot 10 + 9$	$\cdots$	$2 \cdot n + (n-1)$
•	$2 \cdot 0 + 2$	$2 \cdot 1 + 3$	$2 \cdot 2 + 4$	$2 \cdot 3 + 5$	$\cdots$	$2 \cdot 9 + 11$	$\cdots$	$2 \cdot (n-1) + (n+1)$
•	$1 \cdot 2 - 0^2$	$2 \cdot 3 - 1^2$	$3 \cdot 4 - 2^2$	$4 \cdot 5 - 3^2$	$\cdots$	$10 \cdot 11 - 9^2$	$\cdots$	$n \cdot (n+1) - (n-1)^2$

Finding different algebraic expressions for the same pattern, as in this example, is a useful way to introduce algebraic operations and properties which lead to simplifying expressions to reveal their equivalence.

- $2n + (n-1) = 2n + n - 1 = (2n+n) - 1 = 3n - 1$
- $2(n-1) + (n+1) = 2n - 2 + n + 1 = 2n + n - 2 + 1 = 3n - 1$
- $n(n+1) - (n-1)^2 = n^2 + n - (n^2 - 2n + 1) = n^2 + n - n^2 + 2n - 1 = 3n - 1$

Next we will consider another approach to this problem. The following table provides a numeric approach to examining the previous pattern:

Figure # (x)	0	1	2	3	$\cdots$	10	$\cdots$	100
# of dots (y)		2	5	8	$\cdots$	29	$\cdots$	299

The previous pattern can also be extended to functions and lines by studying the graph of the line $y = 3x - 1$ for the x values $1, 2, 3, \cdots$. Note the correspondence of the graph with the previous table.

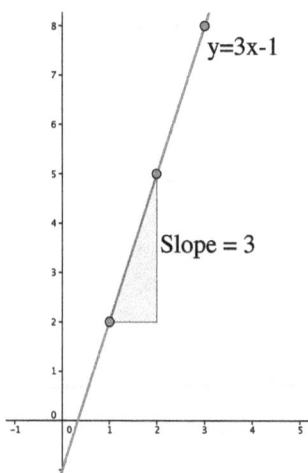

In this example we have illustrated algebraic thinking by connecting different mathematical concepts, using input/output tables, graphing coordinate points, writing equations, and finding slopes and intercepts. Unfortunately, these deeply connected concepts are often taught separately in the K-12 curriculum. Future teachers are urged to make these connections in the classroom in order to create opportunities to develop conceptual understanding.

You can see another example of the use of multiple representations for developing algebraic thinking on page 117.

4.3 INVESTIGATIONS

Investigation 16: Algebraic Expressions

1. Write three number sentences to describe how many dots you see in the figures below. Circle or color parts of the figures to indicate the numbers in your sentences.

2. Using the same three number sentences used in part 1 describe how many dots you see in the following figures.

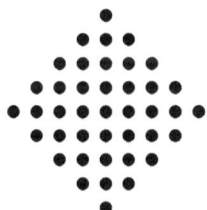

Investigation 17: Dot Patterns

Write a number sentence to describe each figure below, then use the same number sentence to find how many dots will be in figures 4, 10, 100, and figure *n*.

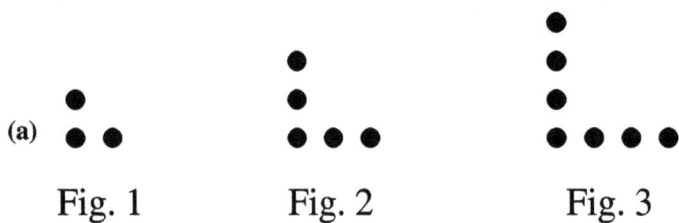

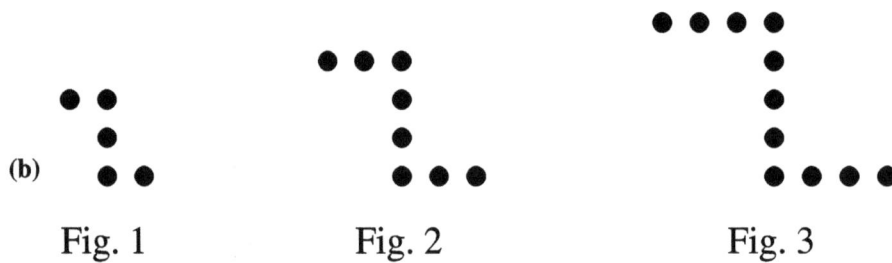

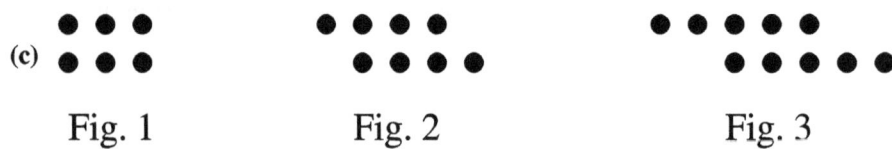

(a)

Fig. 1   Fig. 2   Fig. 3

(b)

Fig. 1   Fig. 2   Fig. 3

(c)

Fig. 1   Fig. 2   Fig. 3

Investigation 18: Square Patterns

Write a number sentence for the number of tiles needed for figures 4, 10, 100, and *n*. Make a table, draw a graph and write an equation for the graph using multiple representations, as discussed in section 4.2 (you can see an example of the use of multiple representations for developing algebraic thinking in page 117).

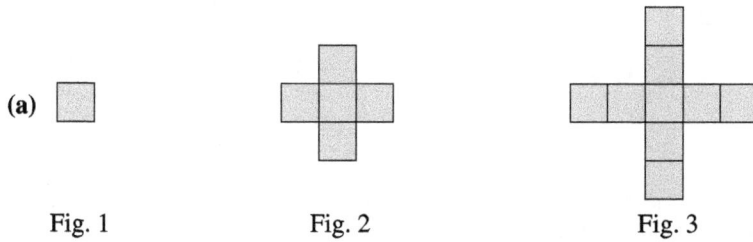

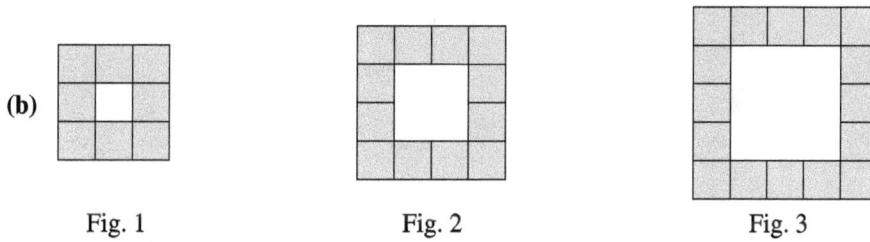

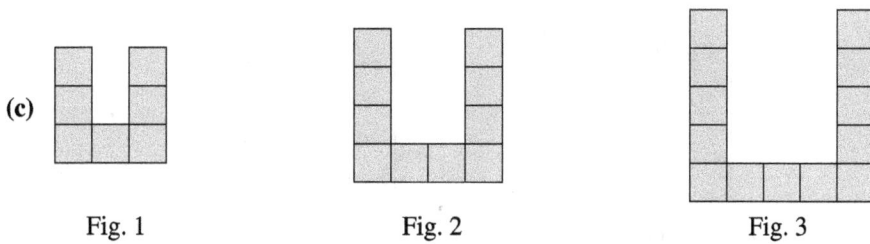

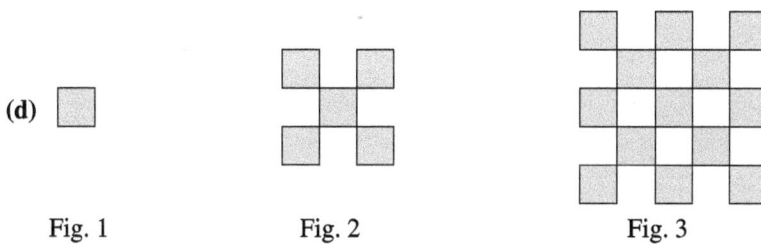

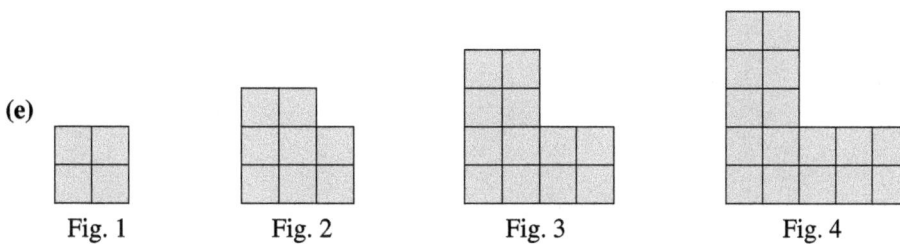

(a)

Fig. 1   Fig. 2   Fig. 3

(b)

Fig. 1   Fig. 2   Fig. 3

(c)

Fig. 1   Fig. 2   Fig. 3

(d)

Fig. 1   Fig. 2   Fig. 3

(e)

Fig. 1   Fig. 2   Fig. 3   Fig. 4

(f)

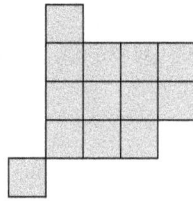
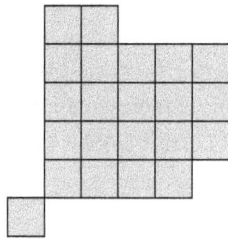
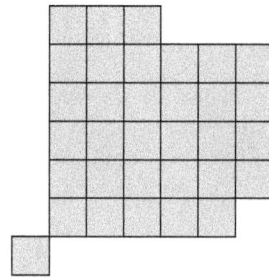

Fig. 1   Fig. 2   Fig. 3

(g)

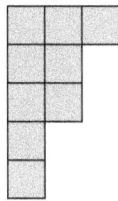
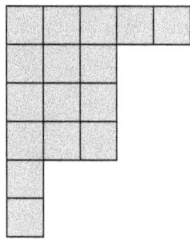
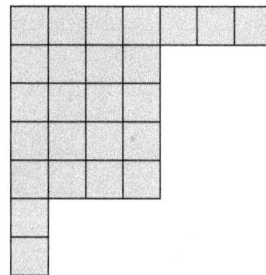

Fig. 1   Fig. 2   Fig. 3

(h)

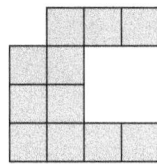
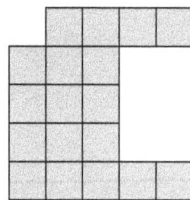
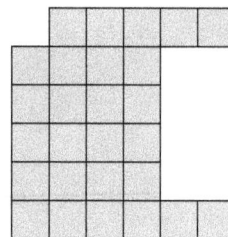

Fig. 1   Fig. 2   Fig. 3

Investigation 19: Staircase Problem

Children were building a staircase with wooden blocks. The first step was build with one block, the second with two stacked blocks, third with three and so forth. Write number sentences for the total number of blocks needed for each staircase, for figure 1, 2, 3, 4, 10, 100 and n.

• Discuss how to find the total number of blocks needed for each of these staircases (try manipulating numbers, manipulating blocks, manipulating picture etc.)

• If you need 210 steps in your staircase how many total blocks do you need?

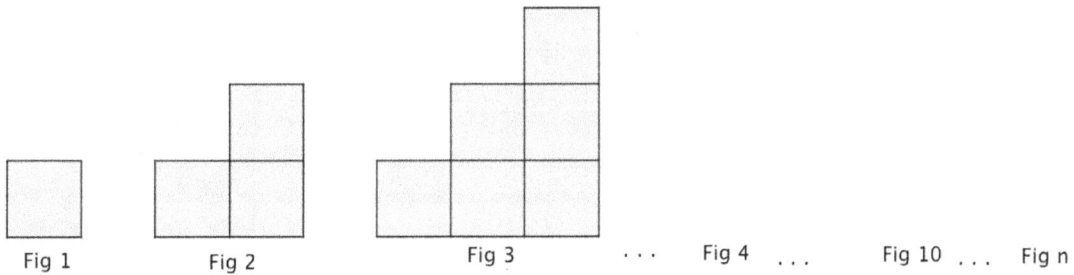

Fig 1   Fig 2   Fig 3   . . .   Fig 4   . . .   Fig 10 . . .   Fig n

Investigation 20: Explorations With Patterns

1. For each of the following sequences of figures:

   - draw the next figure,
   - find the number of squares/triangles in each figure,
   - find the number of squares/triangles in the 50th figure in the sequence,
   - find the area and the perimeter of each of the first five figures and of the 50th figure in the sequence,
   - organize your data in a table as shown in (a).

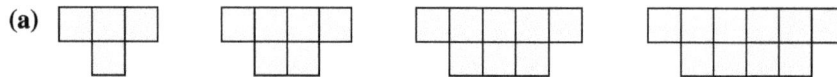

   (a)

Figure in sequence	Number of squares	Area	Perimeter
1st	4	4 square units	10 units
2nd			
3rd			
4th			
5th			
50th			

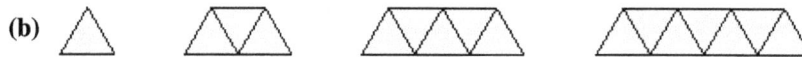

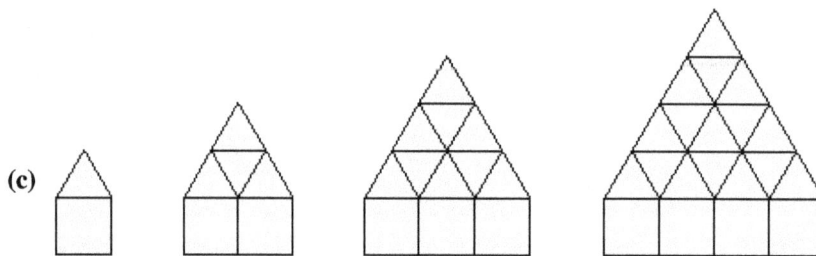

   (b)

   (c)

2. Create your own sequence of figures (using rectangles, triangles, hexagons, or any other shapes you like). Find the number(s) of shapes needed for the 50th figure in your sequence. Can you also find its area and/or perimeter?

Investigation 21: Number Patterns

1. Find the following sums:

   (a) 1   **(b)** 1+3   **(c)** 1+3+5   **(d)** 1+3+5+7

   What do you notice? Do you think the pattern will continue? Why? Calculate

   $$1+3+5+7+\cdots+97+99$$

2. Calculate the following differences:

   (a) $\dfrac{1}{1}-\dfrac{1}{2}$ **(b)** $\dfrac{1}{2}-\dfrac{1}{3}$ **(c)** $\dfrac{1}{3}-\dfrac{1}{4}$ **(d)** $\dfrac{1}{4}-\dfrac{1}{5}$

   What do you notice? What is the value of

   $$\frac{1}{99}-\frac{1}{100}?$$

3. Calculate the following sums (reduce your answers):

   (a) $\dfrac{1}{1\cdot 2}$

   (b) $\dfrac{1}{1\cdot 2}+\dfrac{1}{2\cdot 3}$

   (c) $\dfrac{1}{1\cdot 2}+\dfrac{1}{2\cdot 3}+\dfrac{1}{3\cdot 4}$

   (d) $\dfrac{1}{1\cdot 2}+\dfrac{1}{2\cdot 3}+\dfrac{1}{3\cdot 4}+\dfrac{1}{4\cdot 5}$

   (e) $\dfrac{1}{1\cdot 2}+\dfrac{1}{2\cdot 3}+\dfrac{1}{3\cdot 4}+\dfrac{1}{4\cdot 5}+\dfrac{1}{5\cdot 6}$

   What do you notice? Do you think the pattern will continue? Why? (Hint: use previous problem to write each fraction as a difference, then cancel terms.) What is the value of

   $$\frac{1}{1\cdot 2}+\frac{1}{2\cdot 3}+\frac{1}{3\cdot 4}+\cdots+\frac{1}{98\cdot 99}+\frac{1}{99\cdot 100}?$$

4. The sequence $1,1,2,3,5,8,13,\cdots$, where each number is equal to the sum of two previous numbers, is called the Fibonacci sequence. The numbers in this sequence are called Fibonacci numbers.

   (a) Calculate the next 4 Fibonacci numbers.

   (b) Determine the parity (i.e. whether it is even or odd) of each number in the sequence and describe the pattern. Do you think the pattern will continue? Explain why. (That is, explain how you can be sure that it will.)

5. Calculate the following sums:

(a) 1

(b) $1+2$

(c) $1+2+3$

(d) $1+2+3+4$

(e) $1+2+3+4+5$

(f) 1^3

(g) 1^3+2^3

(h) $1^3+2^3+3^3$

(i) $1^3+2^3+3^3+4^3$

(j) $1^3+2^3+3^3+4^3+5^3$

Compare the two sequences you get. What do you notice? Describe the pattern in words and write a formula. Can you explain why this pattern always holds, no matter how far you go in the sums?

Investigation 22: Area Model

1. Draw the area model and use it to find the answers to the following multiplication tasks:

   (a) $23 \times 34 =$

   (b) $(x+6)(x+3) =$

   (c) $(2x+0.5)(3x+1) =$

   (d) $(a+0.4)(3.2+b) =$

   (e) $(a+b+0.5c)(2a+3c) =$

2. Draw the area model and use it to find the answers to the following division tasks:

   (a) $36 \div 9 =$

   (b) $x^2 \div x =$

   (c) $(x^2+7x) \div (x+7) =$

   (d) $(x^2+7x) \div x =$

   (e) $(x^2+7x+12) \div (x+4) =$

Investigation 23: Making Sense of *xy* Algebra Tiles

The *xy* algebra tiles can be used to illustrate many algebraic concepts such as operations with expressions and solving linear and quadratic equations.

Here is a list of *xy* algebra tiles and their areas.

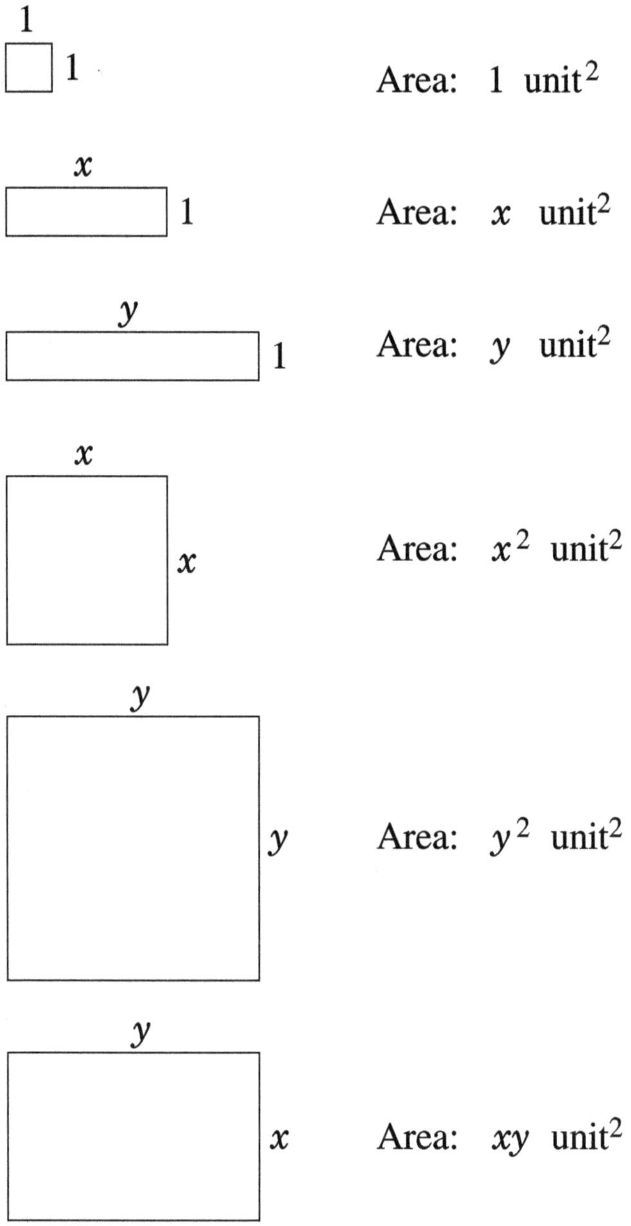

1

Area: 1 unit^2

x

Area: x unit^2

y

Area: y unit^2

x

Area: x^2 unit^2

y

Area: y^2 unit^2

y

Area: xy unit^2

Example. Consider the following figure.

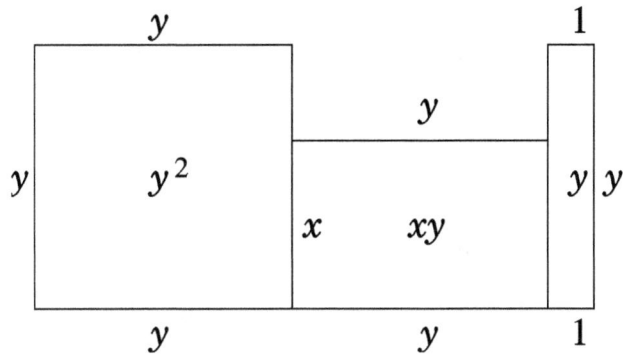

The area of the above figure is $y^2 + xy + y$, and the perimeter is

$$y + y + 1 + y + 1 + (y - x) + y + (y - x) + y + y = 8y - 2x + 2$$

Find the area and the perimeter of each of the following figures.

1.

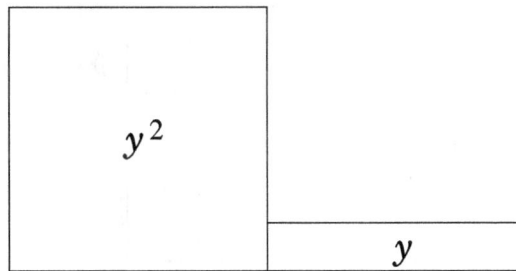

2.

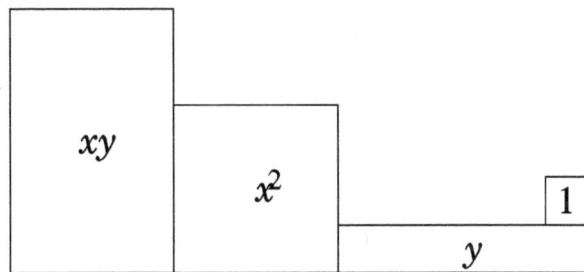

3.

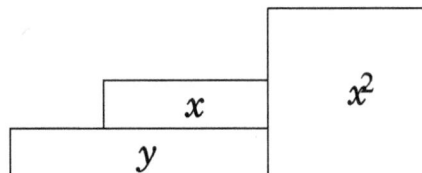

4.

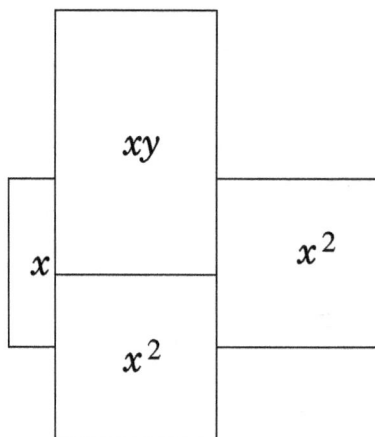

5.

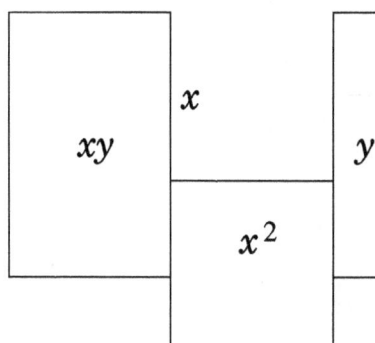

Investigation 24: Addition of Algebraic Expressions Using xy Algebra Tiles

Let's explore how to add integers and expressions using algebra tiles.

Example 1.
We want to calculate $3 + 2$

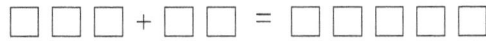

So, $3 + 2 = 5$

Example 2.
Now we want to find $^-3 + 2$

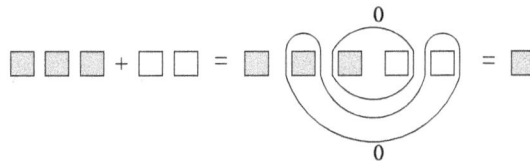

Then, $^-3 + 2 = -1$

Example 3.
Let us calculate $2x + 3 + {^-x}$

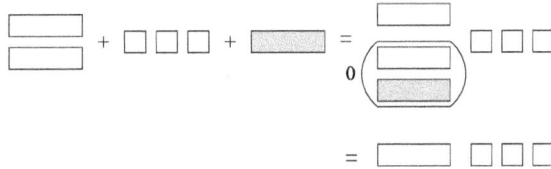

We get that $2x + 3 + {^-x} = x + 3$

Example 4.
We want to simplify $-y + 3x^2 + 2 + 2y + {^-3}$

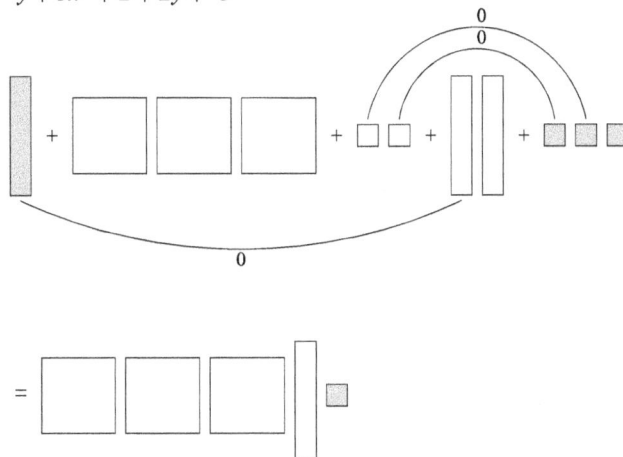

Then, $-y + 3x^2 + 2 + 2y + {^-3} = 3x^2 + y - 1$

Investigation 25: Subtraction of Algebraic Expressions Using xy Algebra Tiles

- Take away model

Example 1.
We want to compute $3 - 2$

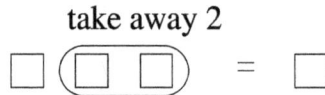

take away 2

It follows that $3 - 2 = 1$

Example 2.
Now, let us find $^-5 - {}^-2$

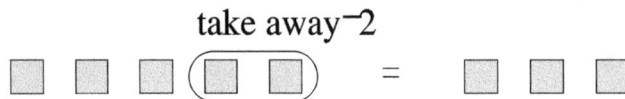

take away $^-2$

We get $^-5 - {}^-2 = {}^-3$

Example 3.
We want to compute $^-3 - 1$

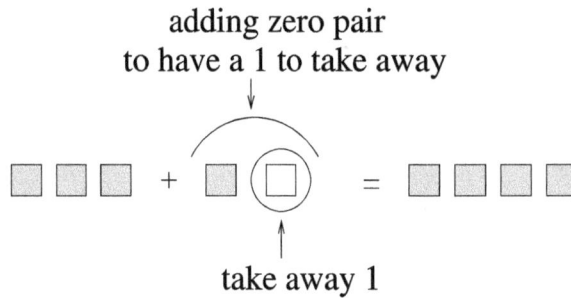

adding zero pair
to have a 1 to take away

take away 1

So, $^-3 - 1 = {}^-4$

Example 4.
We want to compute $2x - {}^-x$

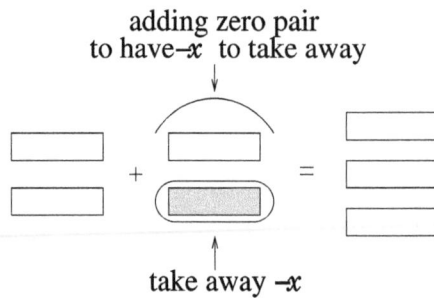

adding zero pair
to have ^-x to take away

take away ^-x

We get $2x - {}^-x = 3x$

Using the take away model, simplify the following:

(a) $2 - 2x - {}^-2$

(b) $2x^2 - {}^-x - 2xy - x$

- Adding the opposite model

Example 1.
$${}^-5 - 7 = {}^-5 + (7 + {}^-7) - 7 = {}^-5 + (7 - 7) + {}^-7 = {}^-5 + {}^-7 = {}^-12$$

Any 'take away' subtraction can be written as adding the opposite:

Example 2.
$$2x - 3x = 2x + {}^-3x = {}^-x$$

Example 3.
$$2x - (x + 2) = 2x + ({}^-x + {}^-2) = 2x + {}^-x + {}^-2 = x + {}^-2 = x - 2$$

Investigation 26: Multiplication of Algebraic Expressions Using xy Algebra Tiles

Multiplying algebraic expressions using algebra tiles is similar to multiplying integers using base 10 tiles. Recall that

- 12×13 can be represented by base 10 tiles as follows:

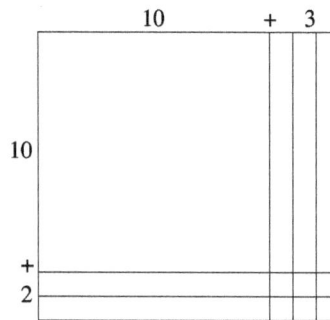

So $12 \times 13 = 100 + 10 + 10 + 10 + 10 + 10 + 6 = 156$.

- $3 \times {}^-2$ means 3 groups of $^-2$:

so $3 \times {}^-2 = -6$.

- $^-3 \times 2$ means the opposite of 3 groups of 2:

3x2   opposite of 3x2

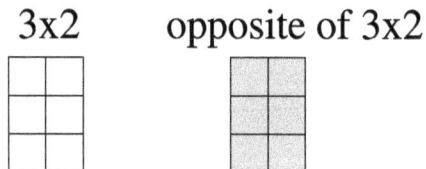

so $^-3 \times 2 = -6$.

- $^-3 \times {}^-2$ means the opposite of 3 groups of $^-2$:

3x$^-$ 2   opposite of 3x 2

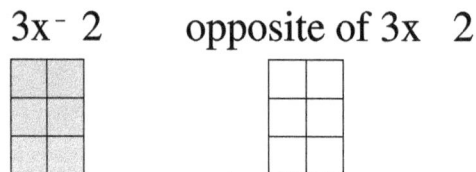

so $^-3 \times {}^-2 = 6$.

Below are a few examples of multiplying algebraic expressions.

Example 1. $(x+2)(x+1)$

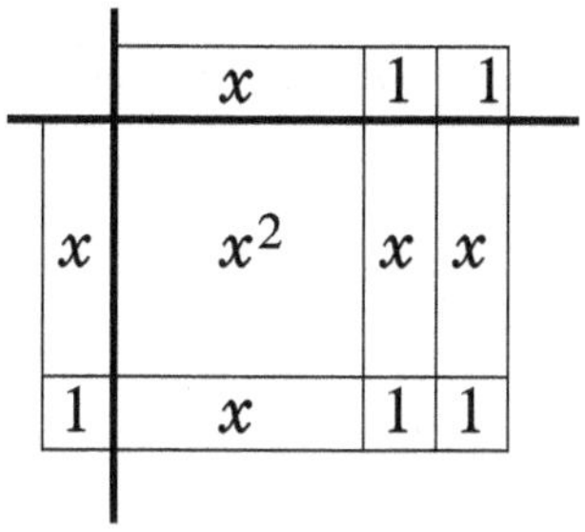

	x	1	1
x	x^2	x	x
1	x	1	1

so $(x+2)(x+1) = x^2 + 3x + 2$

Example 2. $(2x-1)(x-1)$

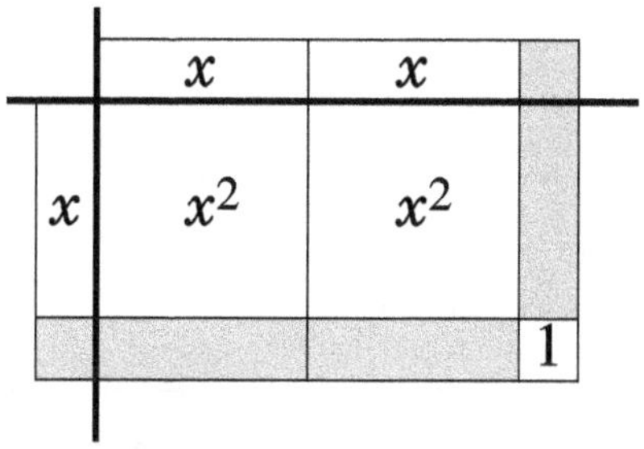

	x	x	
x	x^2	x^2	
			1

so $(2x-1)(x-1) = 2x^2 + {}^-3x + 1 = 2x^2 - 3x + 1$

Example 3. $(2x-1)(2x+1)$

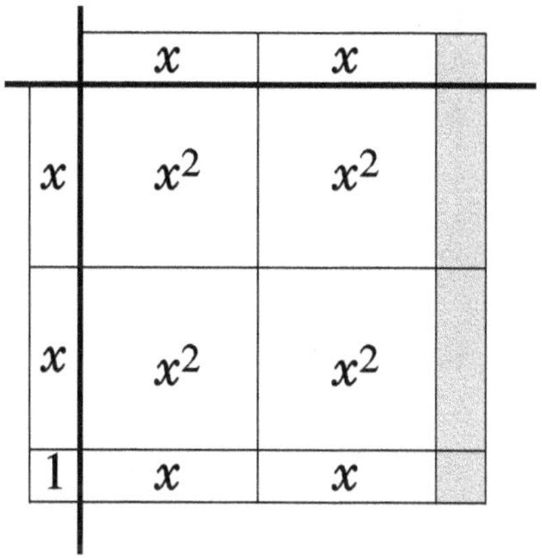

	x	x	
x	x^2	x^2	
x	x^2	x^2	
1	x	x	

so $(2x-1)(2x+1) = 4x^2 + 4x + {}^-4x + {}^-1 = 4x^2 - 1$

Investigation 27: Division of Algebraic Expressions Using xy Algebra Tiles

Division of algebraic expressions can be interpreted as follows: given the area of the rectangle, and one side, find the other side.

For example, $(2x^2 + 3x + 1)/(x + 1) = ?$ means:

$$(x+1)(?) = 2x^2 + 3x + 1,$$

That is,

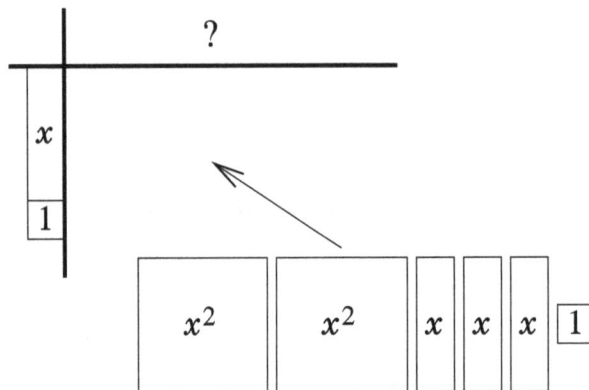

Since

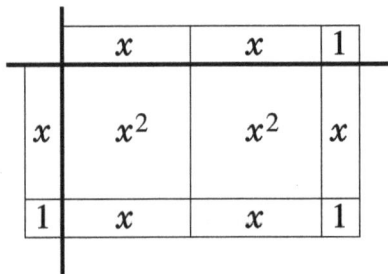

we have $(2x^2 + 3x + 1)/(x + 1) = 2x + 1$.

Try the following division problems using algebra tiles:

(a) $(xy + 3x + y + 3)/(x + 1)$

(b) $(x^2 + xy + 2x + y + 1)/(x + 1)$

Investigation 28: Factoring of Algebraic Expressions Using xy Algebra Tiles

Factoring can be interpreted as: given the area of the rectangle, find it sides.
For example, to factor $y^2 + 2y + 1$, we try to arrange the corresponding tiles into a rectangle:

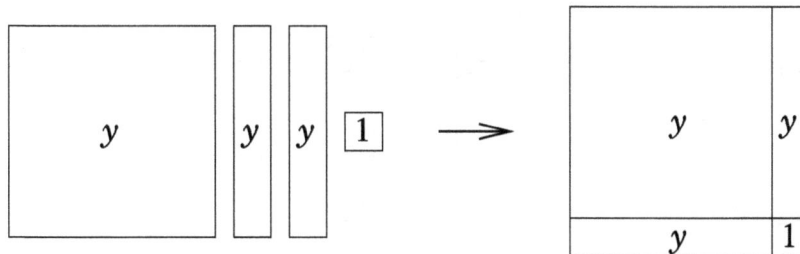

so $y^2 + 2y + 1 = (y+1)(y+1)$.

Factor the following polynomials using algebra tiles:

(a) $x^2 + xy + 2y + 4x + 4$

(b) $xy + 2x + y + 2$

(c) $x^2 + 5x + 6$

(d) $2x^2 + 5x + 3$

(e) $2y^2 - 5y - 3$

(f) $xy + 3x + 2y + 6$

(g) $x^2 - 1$

(h) $4y^2 - 9$

Investigation 29: Solving Linear Equations Using *xy* Algebra Tiles

Example.

$2(x+3) - 4 = 6$

$2(x+3) = 10$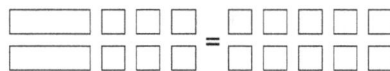

$x+3 = 5$

$x = 2$

Solve the following equations and illustrate your solutions with algebra tiles:

(a) $x - 3 = 4$

(b) $x - 2x = 5 + x$

(c) $2(x+3) = 3 - 2x$

(d) $2(3+x) + 2 = {}^-6$

Investigation 30: Solving Quadratic Equations Using xy Algebra Tiles

Example 1. $x^2 = 4$

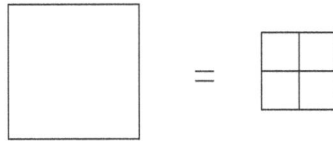

So either $x = 2$ or $x = {}^-2$.

Example 2. $x^2 + 4x + 4 = 1$

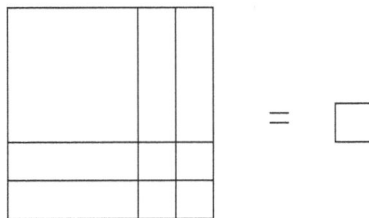

So either $x + 2 = 1$ or $x + 2 = {}^-1$. Thus $x = {}^-1$ or $x = {}^-3$

Example 3. $x^2 + 6x + 1 = 8$

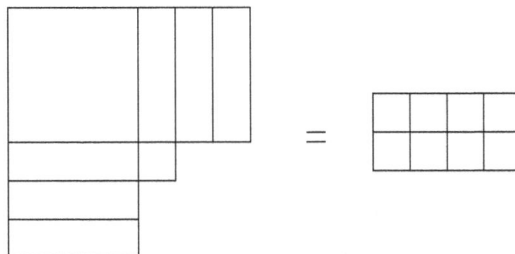

Adding 8 to both sides (so that we can make squares), we have:

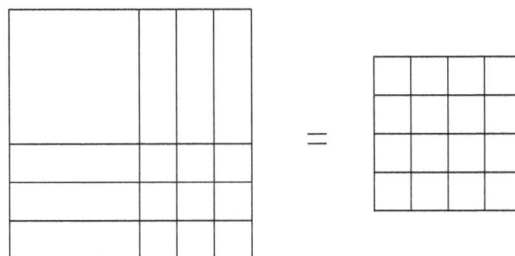

So either $x + 3 = 4$ or $x + 3 = {}^-4$. Thus $x = 1$ or $x = {}^-7$.

Example 4. $x^2 = 7$

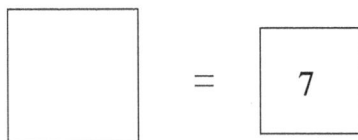

So $x = \sqrt{7}$ or $x = {}^-\sqrt{7}$.

Solve the following equations using algebra tiles. Check your answers using the quadratic formula:

$$x = \frac{-b \pm \sqrt{b^2 - 4ac}}{2a}$$

(a) $x^2 + 6x + 1 = {}^-3$

(b) $x^2 + 4x - 3 = 2$

(c) $(x + 2)^2 = 6$

(d) $x^2 - 6x = {}^-8$

(e) $x^2 + 4x = 5$

(f) $x^2 + 6 = 2x$

(g) $4x^2 + 4x + 1 = 5$

(h) $2x^2 + 2x = 4$

(i) $x^2 + 3x + 2 = 2$

Investigation 31: A Balanced Diet?

Suppose that $\frac{3}{4}$ of a pizza and a bottle of coke filled to $\frac{1}{3}$ are in balance on a scale with $\frac{2}{5}$ of the pizza and a full bottle of coke.

Draw a picture of this situation. Assuming that the weight of the plastic bottle is very small compared to the weight of the coke and can be neglected, use your picture to answer the following questions.

(a) What weighs more, a whole pizza or a full bottle of coke?

(b) Let's use the weight of the full bottle of coke as a weight unit. Express the weight of the pizza in this unit.

(c) Now let the weight of the pizza to be a weight unit. Express the weight of the full bottle of coke in this unit.

(d) Find the ratio: $\dfrac{\text{weight of the pizza}}{\text{weight of the full bottle of coke}}$.

(e) Make a prediction: If you put $\frac{1}{5}$ of a pizza on one side and $\frac{1}{3}$ of the bottle of coke on the other side, will the scale be in balance? If not, what side will be heavier?

Investigation 32: Comparing Solution Methods for Linear Equations

1. Solve each of the following equations in two different ways:
   (i) using algebraic procedures, and
   (ii) using graphs of functions.

   (a) $2(x-30) = 25 - 3x$

   (b) $3 - \dfrac{5x+6}{4} = \dfrac{4+x}{5}$

   (c) $2x + 4 = 2(4+x)$

   (d) $2x + 4 = 2(3+x)$

   (e) $2x + 4 = 2(2+x)$

2. Solve the following system of equations in two different ways:
   (i) using algebraic procedures, and
   (ii) using graphs of functions.

$$2x + 8 = 6y$$
$$3(y - x) = 12$$

Investigation 33: Challenging Questions About Quadratic Equations

1. True or False? Give a reason for your answer.

   (a) The number 20 satisfies the equation $8x - x^2 = 75$.

   (b) The number $\frac{2}{3}$ is a solution of the equation $10x + 3x^2 = 8$.

   (c) The number -3.9 is a solution of the equation $(x - 22.8)(x + 3.9) = 0$.

   (d) The quadratic expression $x^2 - 5x + 6$ can be written as a product of $x - 2$ and $x - 3$.

   (e) The equation $(3x + 2)(4 - x) = 0$ is equivalent to the quadratic equation $3x^2 - 10x - 8 = 0$.

   (f) The equation $3x^2 = 10x - 8$ is equivalent to the quadratic equation $3x^2 - 10x - 8 = 0$.

   (g) The quadratic equation $3x^2 - 10x - 8 = 0$ has two integer solutions.

   (h) The quadratic equation $3x^2 - 10x - 8 = 0$ has two real solutions.

   (i) The quadratic equation $3x^2 + 12x + 12 = 0$ has two integer solutions.

   (j) The quadratic equation $3x^2 - 12x + 12 = 0$ has no negative solution.

2. Show your work:

   (a) Solve for x: $10x - 24 = 0$.

   (b) Solve for x: $x^2 + 10x - 24 = 0$.

   (c) Find all real solutions of the equation $(8x - 5)(2 + 6x) = 0$.

   (d) Find all real solutions of the equation $x^2 = 6$.

   (e) Find all real solutions of the equation $x^2 - 5x = 0$.

   (f) Use the quadratic formula to find all solutions of $3x^2 + 12x + 12 = 0$.

   (g) Use the quadratic formula to solve the equation $3x^2 - x + 24 = 0$.

   (h) Use the quadratic formula to find all solutions of $(x + 3)^2 = 12x$.

4.4 PROBLEM SOLVING

Polya's Corner

A similar problem was given in *Mathematics Without Borders*, an international competition where students do math problems in a foreign language.

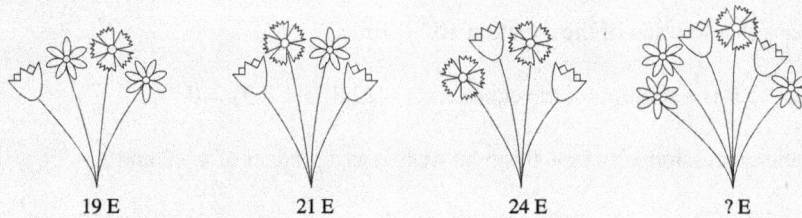

| 19 E | 21 E | 24 E | ? E |

Model Polya's 4 steps:

1. Understand the problem. What is the task here? Explain in your own words. What could the numbers and "*E*" represent in this problem? What kind of number do you expect for the answer?

2. Devise a plan. What strategies could you use to solve this problem?

3. Carry out the plan.

4. Look back. Were your expectations about the type of answer met? Try different strategies. Which method works out the fastest? Do you need to know a specific number associated with each kind of flower for answering the question?

EXERCISES

1. Recall that to solve an equation means to find all solutions, i.e. all numbers x that make the equality true.

   Solve the following equations over the set of real numbers (i.e. find all real solutions of these equations). Show all steps of your solutions. Make sure that you can justify each step (recall that you can add the same number to both sides of an equation, subtract the same number from both sides of an equation, multiply both sides of an equation by the same number, or divide both sides of an equation by the same nonzero number. Be careful with division: if dividing by a variable/expression, remember that division is legal only when the variable/expression is nonzero!)

   (a) $2x + 5 = 19$ (b) $2x + 5 = 4x + 11$

   (c) $3x = 5x$ (d) $3x + \dfrac{x}{2} = 5 - \dfrac{x}{3}$

   (e) $\dfrac{3}{x} = 9$ (f) $x^2 = 16$

   (g) $x^3 = -27$ (h) $(x - 3)(x + 5) = 0$

   (i) $x^2 + 2x - 35 = 0$ (j) $2x^2 + 7x - 4 = 0$

   (k) $3x^2 - 4x - 5 = 0$ (l) $2x^2 - 5x + 7 = 0$

   (m) $(x + 2)(x - 5) = 6$ (n) $x(x - 3) = (2x + 1)(x + 2) - 8$

2. In a small town, three children deliver all the newspapers. Abby delivers 3 times as many papers as Bob, and Connie delivers 13 more than Abby. If the three children deliver a total of 496 papers, how many papers does each deliver?

3. The formula for converting degrees Celsius (C) to degrees Fahrenheit (F) is

$$F = \frac{9}{5}C + 32.$$

   Your European friend asks you how warm it is now in Fresno. Your outdoor thermometer shows 80 degrees Fahrenheit. How many degrees Celsius is it?

4. Two silk butterflies and a silk rose cost $18. One silk butterfly and a silk rose cost $11. What is the cost of each?

5. A teacher instructed her class as follows: Take any number and add 15 to it. Now multiply that sum by 4. Next subtract 8 and divide the difference by 4. Now subtract 12 from the quotient and tell me the answer, I will tell you the original number. Analyze the instructions to see how the teacher was able to determine the original number.

6. Make up your own procedure similar to that in the previous problem. Test it by asking your group members to follow the steps you give them and tell you the result.

7. For an event at school, 812 tickets were sold for a total of $1912. If students paid $2 per ticket and non-students paid $3 per ticket, how many student tickets were sold?

8. Consider the following sequence of figures.

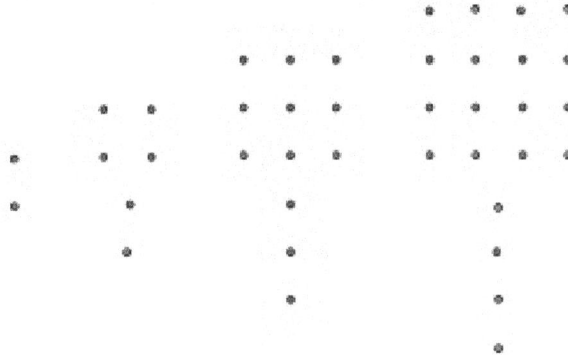

Fig. 1   Fig. 2   Fig. 3   Fig. 4

   (a) Write a formula for the number of dots in the nth figure.

   (b) Using your formula find the number of dots in the 60th term.

9. Use algebra tiles to perform the following multiplications.

   (a) $(2x+3)(x+4)$ (b) $(x-2)(x+2)$

   (c) $(x-3)(x+3)$ (d) $(x+3)^2$

   (e) $(x+1)^2$ (f) $(x-2)^2$

   (g) $(x-3)^2$ (h) $(x+3)(x-2)$

   (i) $(x-1)(x+2)$ (j) $(x+3)(x+4)$

   (k) $(x-4)(2x+3)$ (l) $(x-4)(2x-1)$

   (m) $(x+1)(2x-5)$ (n) $(x+1)(x-1)$

10. With the techniques/ideas used in the previous problem, get a formula for

   (a) $(x-y)(x+y)$ (b) $(x+y)^2$

   (c) $(x-y)^2$ (d) $(x+a)(x+b)$

11. Factor the following polynomials using algebra tiles.

   (a) $x^2 - 9$ (b) $x^2 + 2x + 1$

   (c) $x^2 - 6x + 9$ (d) $x^2 + 7x + 12$

   (e) $x^2 + x - 2$ (f) $x^2 - x - 2$

   (g) $x^2 + 5x + 6$ (h) $2x^2 + 3x - 2$

   (i) $3x^2 - 4x - 4$ (j) $2x^2 - 11x + 12$

   (k) $2x^2 - 7x + 5$ (l) $x^2 - 2$

12. Solve the following quadratic equations using algebra tiles (completing the square or factoring) and also using the quadratic formula.

   (a) $x^2 + 4x - 12 = 0$ (b) $x^2 + 7x + 12 = 0$

   (c) $x^2 + x = \frac{5}{4}$ (d) $x^2 - 2x = 3$

   (e) $x^2 - 5x + 4 = 16 - 9x$ (f) $x^2 - 5x + 4 = -8 - 12x$

   (g) $x^2 + 10x + 6 = 3x - 6$ (h) $x^2 - 1 = 0$

13. Find at least two different solutions for the following problem. One of them must be by setting and solving an equation.
    A farmer calculates that out of every 100 seeds of corn he plants, he harvests 84 earn of corn. If he wants to harvest 7200 ears of corn, how many seeds must he plant?

14. Find at least two different solutions for the following problem. One of them must be by setting and solving an equation.
    A photograph measuring 3 inches by $2\frac{1}{2}$ inches is to be enlarged so that the smaller side, when enlarged, will be 8 inches. How long will the enlarged longer side be?

15. Find at least two different solutions for the following problem. One of them must be by setting and solving an equation.
    Karl wants to fertilize his 6 acres. If it takes $8\frac{2}{3}$ bags of fertilizer for each acre, how much fertilizer does Karl need to buy?

16. Find at least two different solutions for the following problem.
    A drink and a sandwich together cost $5.95. The sandwich costs $2.45 more than the drink. How much does the sandwich cost?

17. Find at least two different solutions for the following problem.
    A rectangular field is to be enclosed with 1180 feet of fencing. If the length of the field is 50 feet longer than the width, then how wide is the field?

18. Find at least two different solutions for the following problem.
    Robert has a rectangular garden with area 280 ft^2 and a side with length $10\frac{2}{3}$ feet. What is the length of the other side?

19. Find at least two different solutions for the following problem.

    A group of friends want to mow the lawn at their yard (in the shape of a square), they have decided to each take care of a strip of grass. If the side of the square is $4\frac{1}{3}$ yards and each will work on a strip with side of $\frac{2}{3}$ yards, then how many friends will work on the yard?

20. Find at least two different solutions for the following problem.

    A rectangular pool has perimeter 300 ft^2. If one of the sides is 3 feet larger than the other, find the lengths of both sides.

21. Evaluate for $x = 2$: $\dfrac{5(2x+3)+3}{5+3}$.

22. Solve the following equation for x. $\dfrac{3x+3}{3} + 3 = 20$.

    Draw pictures parallel with the steps of solving the equation to explain what are you doing.

CHAPTER 4 REFLECTIONS

1. Explain how algebra tiles can be used in the elementary school classroom to develop algebraic thinking. Discuss advantages and difficulties.

2. Explain in your own words the meaning of terms 'variables', 'expressions' and 'equations'.

3. Explain what you mean by solving an equation. Explain what important tools/ideas you should keep in mind when you solve an equation.

Chapter 5
Geometric Thinking

5.1 GEOMETRIC THINKING AND THE VAN HIELE LEVELS

Although this text concludes with the topic of geometry, geometric concepts have already played a substantial underlying role in the development of previous chapters' visual models. Many students might describe themselves as visual learners; however, geometric thinking can raise some of the most difficult challenges for them. One reason for difficulties with geometry is related to the blind use of formulas without reasoning and understanding, since for students "not sufficiently grounded in basic geometric concepts and relations to reinvent Euclidean geometry - memorization may be their only recourse" (Burger & Shaughnessy, 1986, p.46).

The van Hiele Theory of Geometric Thought

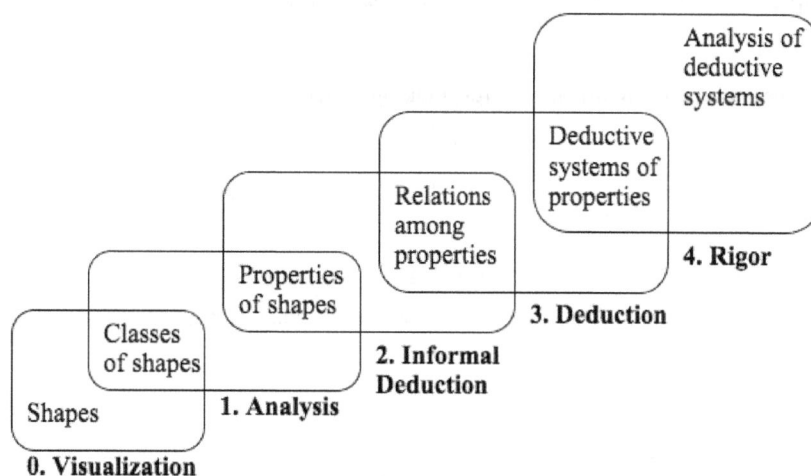

The van Hiele levels, seen in the figure above, describe a progression of phases that students are expected to go through in their development of geometric thinking. The following list relates important ideas, to be kept in mind for the preservice teacher, that have resulted from research involving students and the van Hiele levels (McAnelly, 2011):

Key van Hiele-Level Research Findings

1. Progress from one level to the next is more dependent on educational experiences than age.
2. Levels cannot be skipped.
3. A student may be at different levels in different geometric topics.
4. Properties are intrinsic at one level and become extrinsic at the next level.
5. A teacher who is reasoning at one level speaks a different 'language' from a student at a lower level, thus hampering student understanding.

With this in mind, you, as a prospective teacher, need to reflect on all eight of the Common Core standards for mathematical practice, as each, in the vast context of geometry, has significant relevance.

The 8 Standards for Mathematical Practice in the Context of Geometry

1. Make sense of problems and persevere in solving them.

   • Problem solving with geometry; Polya's 4 steps, diagrams, labeling, geometric definitions; 2D & 3D reasoning.

2. Reason abstractly and quantitatively.

   • Deducing geometric attributes beginning with hands-on activities leading to progressively abstract synthesis of geometric concepts as described by the van Hiele levels.

3. Construct viable arguments and critique the reasoning of others.

   • Deductive and inductive thinking; formal proofs; multiple explanations and viewpoints which lead to flexible thinking and more powerful problem solving.

4. Model with mathematics.

   • Using geometry to model applications for real-world problems.

5. Use appropriate tools strategically.

   • Use of compass and straightedge for constructions; Geoboards, visual models.

6. Attend to precision.

   • Precision in measurement, significant figures; making coherent logical arguments and proofs.

7. Look for and make use of structure.

   • Use of geometric structure to simplify or understand problems; use of structure to make logical arguments and proofs.

8. Look for and express regularity in repeated reasoning.

   • Development of geometric thinking from the concrete to abstract formal systems, consistent with the van Hiele levels described above.

5.2 UNDERSTANDING GEOMETRY THROUGH MEASUREMENT

The four main areas of study in K-8 geometry are length, area, volume and surface area. Recall that the first phase of the van Hiele levels involves understanding the attributes and definitions of geometric objects; however, important to progressing to the level of analysis is the assigning of numerical values, or *measurements*, to component-parts of geometric objects. Definitions of quantities such as perimeter, area, volume and surface area are all closely related to their units of measurement. For this reason, understanding the units of measurement for these common geometric quantities helps pave the way for later study of geometric properties.

Use of Dimensional Analysis in Geometry

When working with units of measurement, it is useful to know that, in terms of exponent rules, they can be treated much like algebraic variables. The key to dimensional analysis is understanding the fractional nature of the conversion factor concept.

For example, since $12\,in = 1\,ft$, then when we can form either the fraction, $\frac{12\,in}{1\,ft}$ or the fraction $\frac{1\,ft}{12\,in}$, which both equal 1 since anytime we divide a nonzero number by itself, it is 1. Thinking of conversion factors in this way is a good strategy for knowing which way to use a factor, keeping in mind what unit one wants to convert to, and what unit one wants to cancel (convert from).

Example 5.1. Convert 24 *in* into feet .

Since we want to convert 24 *in* into feet, this tells us we want to multiply by $\frac{1\ ft}{12\ in}$, because then the *inches* will cancel out, like variables, and we will be left with what we want, which is *feet*.

$$24\ in \cdot \frac{1\ ft}{12\ in} = \frac{24 \cdot 1\ in \cdot ft}{12\ in} = 2\ ft$$

Length - measured in units such as inches (*in*), feet (*ft*), miles, centimeters (*cm*), etc. Often a generic unit (1 *u* is used, such as in the diagram below, describing 1.4 *u* or $1\frac{2}{5}$ *u*.

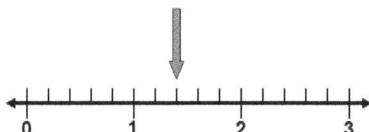

Measurement of perimeter is a common use of length in geometry. The primary two categories perimeter is studied at the K-8 levels are polygons and circular shapes. In Investigations 34 and 35, the perimeter of polygons is studied by using the Pythagorean theorem. Investigation 37 is used for the development of the circumference of a circle (perimeter) and the meaning of the number π.

Area - measured in unit squares and defined as the number of unit squares covering a region (with no overlap). Examples of area units include square inches (in^2), square feet (ft^2), square miles, square centimeters (cm^2), etc. For generic square units, (u^2) is used, as in the unit square below, made from two 1 *u* segments forming the sides of a square, giving an area of $1\ u \cdot 1\ u = 1 \cdot 1 \cdot u \cdot u = 1\ u^2$.

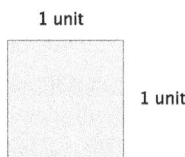

1 unit

1 unit

For rectangular shapes, the amount of unit squares (possibly fractional) can easily fit into the shape to compute area; hence, the study of area begins with the use of rectangles and rearrangement methods to find areas with unit squares. Based on rectangular area, the areas of more complicated shapes such as parallelograms, triangles and trapezoids can be derived (Investigation 42).

Volume - measured in unit cubes and is defined as the number of unit cubes filling an object (see figure below), with no overlap. Examples of volume units are cubic inches (in^3), cubic feet (ft^3), cubic meters (m^3), etc...

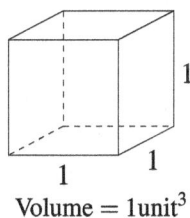

Volume = 1unit^3

The basic three-dimensional geometric objects in the K-8 curricula can be divided into three categories: prism-type objects, pyramid-type objects and spherical objects. Identifying characteristics of prisms helps to see that the number of cubes that fit into the base layer is the same for all other layers. This can contribute to the development of the volume concept for any prism-type object, including cylinders (Investigation 44).

To find the volume of pyramid-type objects, it is important to discover the relationship between prisms and their corresponding pyramids in order to develop the volume concept for pyramid-type objects. Also, power solid manipulatives are very useful tools for these volume investigations. In particular, finding the volume of a sphere can be done by finding the relationship between cylinders and semi-spheres (Investigation 45).

Surface Area - the area of the surface of an object, measured in square units, the same as area. Instead of focusing on formulas for the surface area of different objects, learning to develop the skills to draw two-dimensional net diagrams for three dimensional objects is a good way to calculate the surface area in general, as seen in the figure below depicting the net of a cube having sides of three inches (Investigation 46).

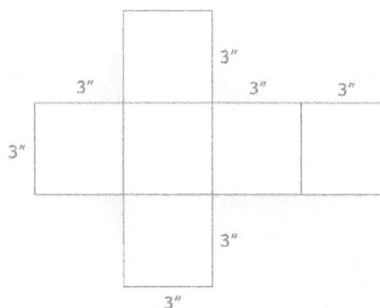

Finding the surface area of cones is more complicated than other objects. Based on techniques for finding the area of a circle, breaking cones into simpler known shapes helps students apply previously developed ideas on area (Investigation 47).

5.3 THE PYTHAGOREAN THEOREM

The Pythagorean theorem is a good place to start for developing geometric thinking because it applies for the most fundamental of triangles, the *right triangle*. Recall that a right triangle is a 3-sided polygon with a 90-degree angle. The hypotenuse of a right triangle is defined as the side opposite the right angle (can there be more than one right angle in a triangle?). Assigning the length of the two legs (call them a and b) and the length of the hypotenuse (call it c), then the Pythagorean theorem says the sides of any right triangle are related by the formula:

$$a^2 + b^2 = c^2$$

Note how the diagram below illustrates a typical example of the Pythagorean theorem since the sum of the areas of the squares of the small sides is equal to the area of the square formed by the hypotenuse, that is, $9 + 16 = 25$. It is important, that when teaching, though, to not subject students *only* to the standard usual examples, such as the $3 - 4 - 5$ right triangle here. Students need to understand that the Pythagorean theorem applies to all right triangles, such as a right triangle having side lengths 1, 1 and $\sqrt{2}$; hence, it is important to convey to students the value of a general proof, which shows the theorem works for *all* cases of right triangles. This geometric interpretation of the Pythagorean Theorem is used for many of its proofs, see Investigation 40.

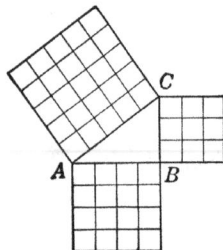

It is always important to keep in mind that when teaching, it is never a good idea to blindly hand down formulas to your students without contributing to the provision of several progressive layers of conceptual justifications, ranging from the concrete (early grades) to the more abstract algebraic and geometric explanations (secondary grades). It is not necessarily your job, as a primary grade teacher, to justify *all* of the layers which provide understanding to a formula; but, it is very important for you to do a thorough job explaining one or two layers at the appropriate level so that advancing students will have firm foundations in which to build on for the more advanced viewpoints they will encounter at the secondary level. Quite simply, understand yourself the formulas you need to teach... and then teach what you understand in as many multiple viewpoints as possible, given time constraints.

5.4 INVESTIGATIONS

Investigation 34: Using Geoboards to Develop Perimeter and Area Concepts

1. Using Geoboards construct a rectangle whose area is 12 square units in three different ways.

2. Construct a rectangle on the Geoboard whose perimeter is 24 units is three different ways

3. Construct geometric shapes on the Geoboard whose area is 18 square units and perimeter is 18, 21, and 24 units long.

4. Find the area of the following geometric shapes in at least 2 different ways.

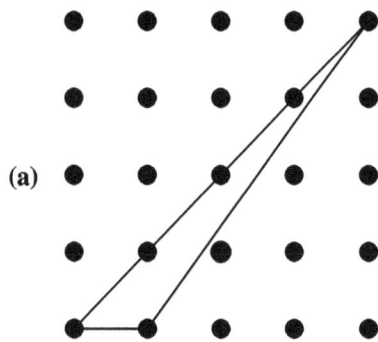

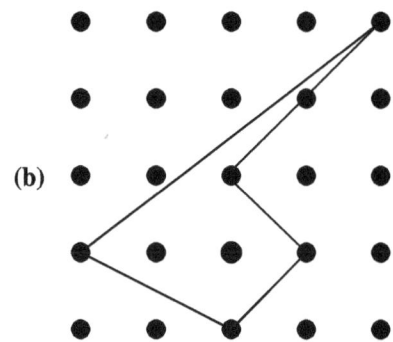

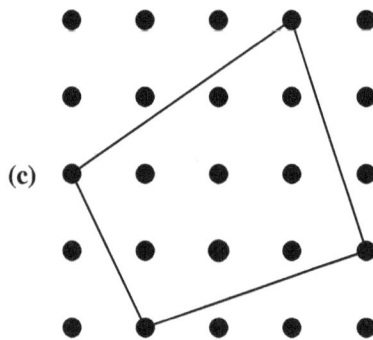

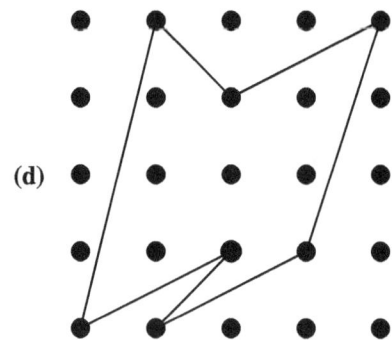

Investigation 35: Exact Geoboard Perimeters

Recalling the Geoboard diagrams of chapter 1, as well as the definition that *perimeter* is the total distance around the outside of a two-dimensional figure, using the Pythagorean theorem allows us to find the exact perimeters of Geoboard figures. Before giving an example of an exact perimeter Geoboard problem, the topic of reducing radicals is prudent to review:

Reducing Radicals

Based on the property of exponents: $\sqrt{a \cdot b} = \sqrt{a} \cdot \sqrt{b}$; radicals like $\sqrt{75}$ can be simplified since the number 75 has a perfect square as a factor, namely 25. Kind of like reducing a fraction, the idea of reducing radicals is to square root out all the *pairs* of factors of the original number, that represented perfect squares, as in the following:

$$\sqrt{75} = \sqrt{25 \cdot 3} = \sqrt{25} \cdot \sqrt{3} = 5\sqrt{3}$$

Example 5.2. Find the exact simplified perimeter of the following 4-gon, or quadrilateral, outlined in solid black (the shortest distance between pegs is 1 unit, or 1 u):

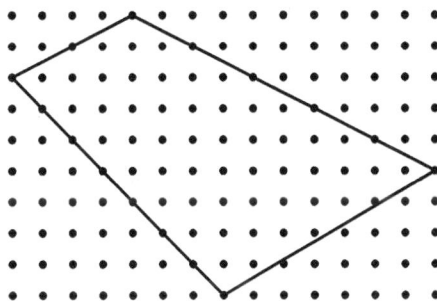

Working clockwise from the top, four right triangles can be formed with the edges of the polygon acting as the hypotenuse for each right triangle. If we can find the length of the hypotenuse of each of four right triangles, then we can get the exact perimeter.

$$
\begin{aligned}
RT\#1: \quad & 2^2 + 4^2 = h_1^2 \longrightarrow h_1 = \sqrt{20} = \sqrt{4 \cdot 5} = \sqrt{4}\sqrt{5} = 2\sqrt{5} \\
RT\#2: \quad & 7^2 + 7^2 = h_2^2 \longrightarrow h_2 = \sqrt{98} = \sqrt{2 \cdot 49} = \sqrt{2}\sqrt{49} = 7\sqrt{2} \\
RT\#3: \quad & 4^2 + 7^2 = h_3^2 \longrightarrow h_3 = \sqrt{65} = \sqrt{5 \cdot 13} = \sqrt{65} \\
RT\#3: \quad & 10^2 + 5^2 = h_4^2 \longrightarrow h_4 = \sqrt{125} = \sqrt{5 \cdot 25} = \sqrt{5}\sqrt{25} = 5\sqrt{5}
\end{aligned}
$$

So, the perimeter of the quadrilateral is

$$P = 2\sqrt{5} + 7\sqrt{2} + \sqrt{65} + 5\sqrt{5} = 7\sqrt{2} + 7\sqrt{5} + \sqrt{65} \ u$$

(a) Find the exact perimeters for the Geoboard figures in problem 4 of Investigation 34. Keep in mind to collect all like radical terms.

(b) Find the exact perimeters for the Geoboard figures in Exercise 7 in chapter 1. Keep in mind to collect all like radical terms.

Investigation 36: Explorations with Pattern Blocks

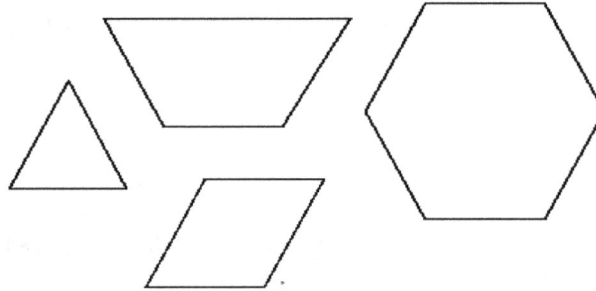

(a) If the area of the triangle is 1 square unit, what are the areas of the other three figures?

(b) If the area of the hexagon is 1 square unit, what are the areas of the other three figures?

(c) If the side of the triangle is 1 unit, what are the areas of all four figures?

(d) If the area of the given hexagon is 1 square unit, what is the area of the hexagon whose each side is twice longer? What if each side is three times longer? Cover such hexagons with the above figures and verify.

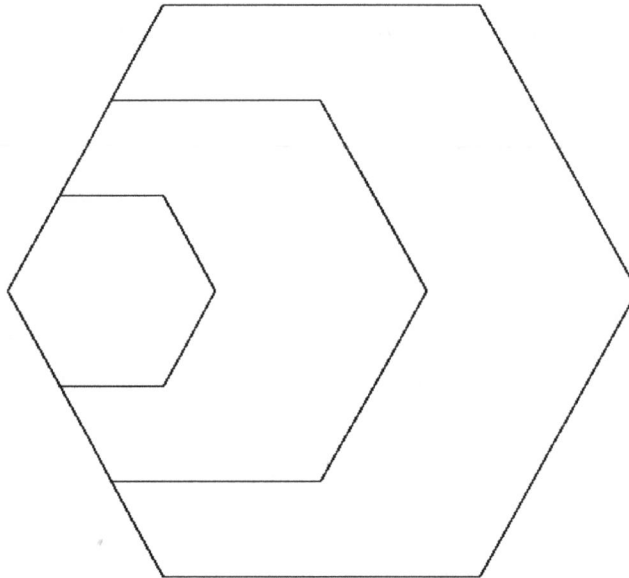

Investigation 37: Circumference of a Circle

1. Find circular objects or draw several circles using a string or a compass.

2. Measure the diameter, radius, circumference of the circles using a string, ruler or a measuring tape.

3. Make a table to find out how these three measures are related to each other.

Circle	radius (r)	diameter (d)	circumference (c)	$\dfrac{d}{r}$	$\dfrac{c}{d}$	$\dfrac{c}{r}$
1.						
2.						
3.						
4.						

4. Graph or calculate the ratios. What do you get?

5. How does this activity help you to find the circumference of any given circle?

Investigation 38: Understanding the Pythagorean Theorem

The Pythagorean theorem is one of the most famous results in mathematics. It is taught in 7th grade, and it appears again in high school standards. Given the beauty of this result, many mathematicians have spent thousands of hours trying to find new and 'better' proofs for this result.

The Pythagorean theorem says that in a right triangle, like the one below,

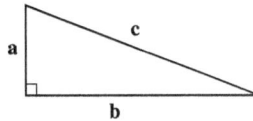

the sum of the squares of the legs is equal to the square of the hypotenuse. That is

$$a^2 + b^2 = c^2$$

1. How can we sort triangles according to their angles?

2. How can we sort triangles according to their sides?

3. What makes a triangle right?

4. How many right angles can a triangle have?

5. Mathematical theorems, most of the times, are written in the form '*if A then B*', where *A* stands for the condition(s) and *B* for the consequence(s).
   Write the Pythagorean theorem in the form *if A then B*.

6. The converse of a statement '*if A then B*' is '*if B then A*'.
   Write down the converse of the Pythagorean theorem.

7. What do you think? Is the converse of the Pythagorean theorem also true?

Investigation 39: Interpretations of the Pythagorean Theorem

Is the geometric interpretation of Pythagorean Theorem true for the following figures?

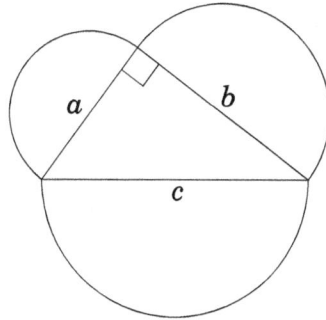

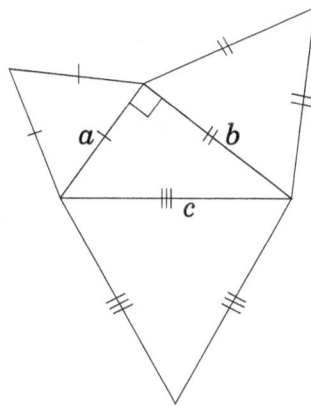

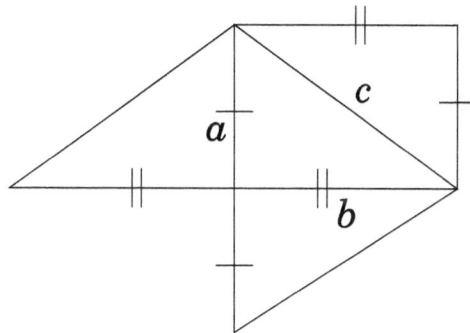

Investigation 40: Proofs of the Pythagorean Theorem

Many proofs of the Pythagorean theorem require finding the area of certain diagrams in two different ways. An easy way to understand where this theorem comes from is by using a little algebra.

For example, the area of the figure below

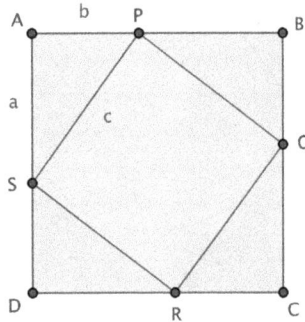

may be found using the side lengths of the square ABCD as $(a+b)^2$. On the other hand, the area of the same figure may be found by using the square PQRS and the four right triangles to be

$$c^2 + 4\left(\frac{1}{2}ab\right)$$

Since both methods give the area of the same figure then the two expressions obtained should be equal to each other. Therefore,

$$(a+b)^2 = c^2 + 4\left(\frac{1}{2}ab\right)$$

After simplifying you will get $a^2 + b^2 = c^2$.

1. Prove the Pythagorean theorem using the following figures
   (a)

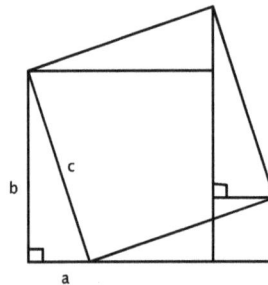

(b)

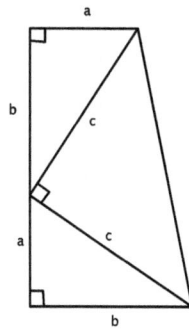

(c)

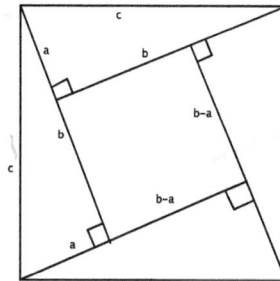

2. Use the following figures to give a proof of the Pythagorean theorem that does not use algebra, such as the ones given above.

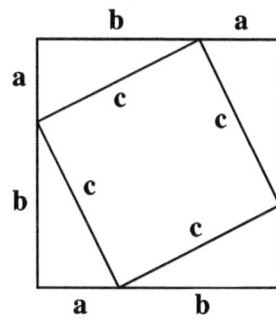
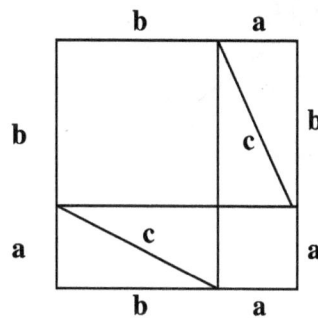

Investigation 41: Applications of the Pythagorean Theorem

Find the value of x in the following figures.

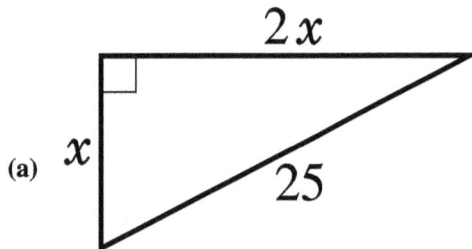

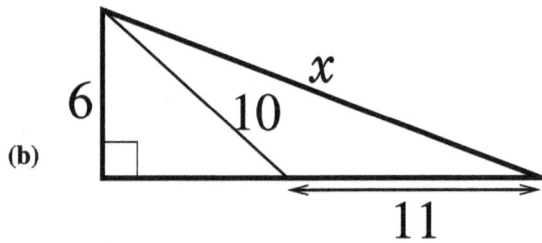

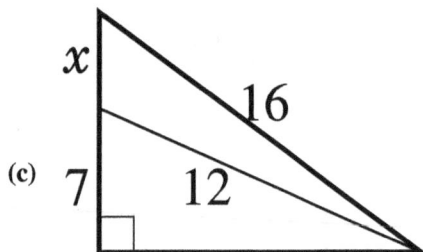

(a)

$2x$

x

25

(b)

6

10

x

11

(c)

x

16

7

12

Investigation 42: Developing Measurement Concepts

The area of a region is measured by the number of non-overlapping square units that can cover the region. These unit squares can be cut and re-organized if necessary.

In order to investigate the process of finding the area of a region we will study how to do this by starting with a simple region and progressively working through more complex ones.

1. Can you explain why the area of the rectangle below is 3 *units* $\times 4$ *units* $= 12$ *unit*^{2}?

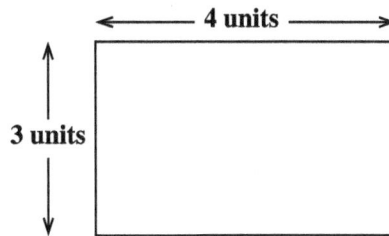

2. Find the area of the following figure in many different ways.

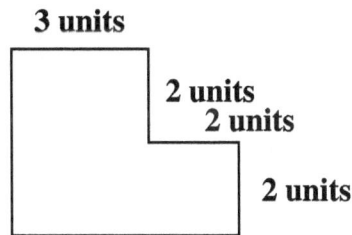

3. Now we would like to find the area of the parallelogram below. Why is its area 3 *units* $\times 4$ *units* $= 12$ *unit*^{2}?

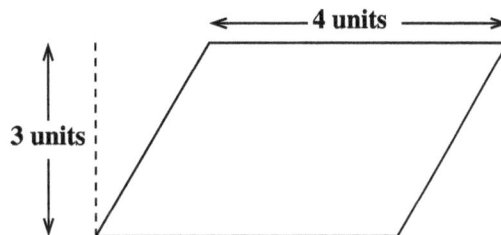

4. Can you use the parallelogram above to find the area of the triangle below?

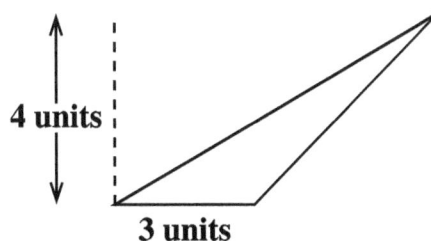

5. The simpler regions have been looked at and understood by now. We should move on to the next level of difficulty: A trapezoid.

   Find the area of the following trapezoid. Explain why its area is 30 $unit^2$.

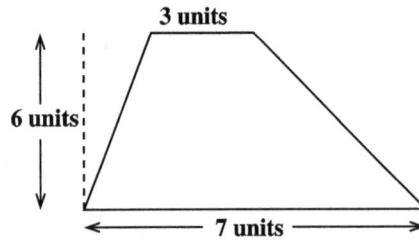

6. We now know about many different shapes. Find the area of the trapezoid above in at least three different ways.

Investigation 43: Area of a Circle

The following diagrams are often used to explain the development of the area of a circle.

Explain how you can obtain that the area of a circle with radius r is πr^2 using the following diagrams

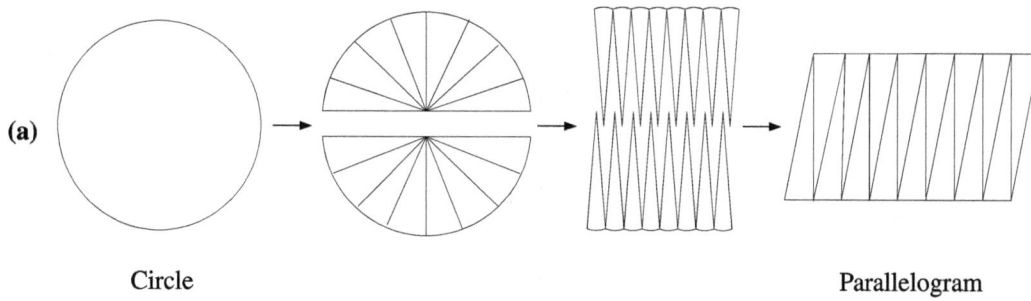

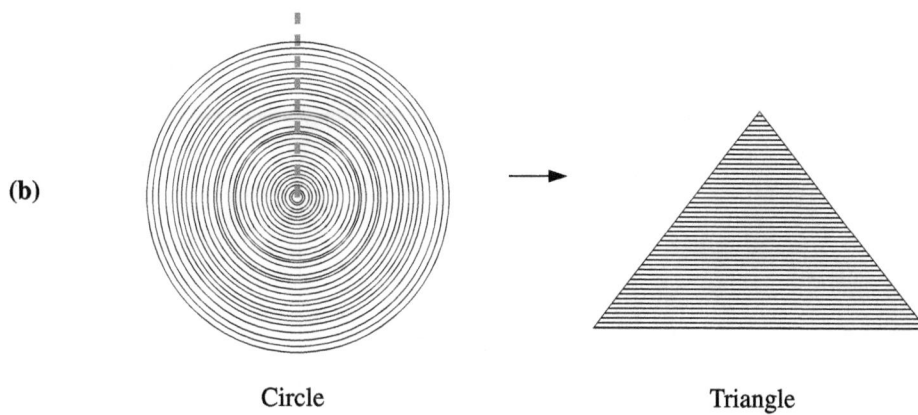

(a)

Circle

Parallelogram

(b)

Circle

Triangle

Investigation 44: Volume

1. Why is the volume of this solid 7 cubic units?

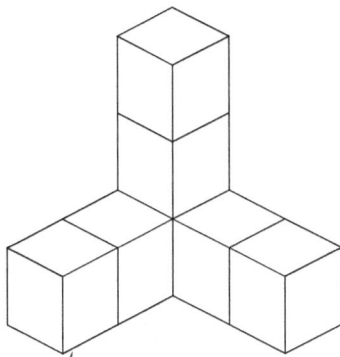

2. Why is the volume of the this solid (rectangular prism) 24 cubic units?

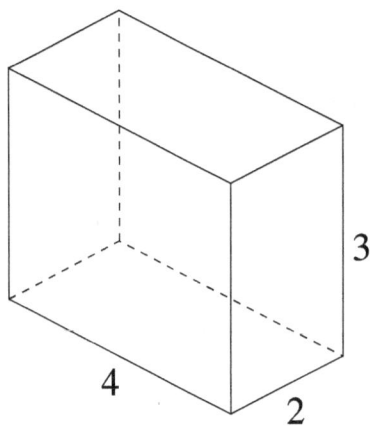

3. Why is the volume of this cylinder 12π cubic units?

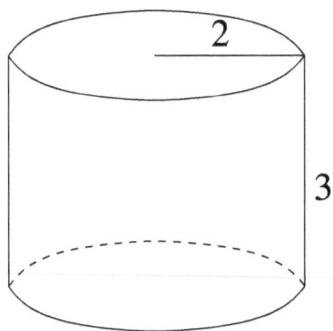

Investigation 45: Volumes of Power Solids

1. Separate the power solids into three different groups. Identify the characteristics of each group.

2. Using the shortest length of all power solids as 1 unit, find the volume of the cylinder and two prism shapes (cube and triangular prism).

3. Using water or sand as a measuring media, compare the volumes of the following pairs to find the volume of pyramids, cones, and spheres.

   (a) cube and square pyramid

   (b) triangular prism and triangular pyramid

   (c) cylinder and cone

   (d) cylinder and semi-sphere

Investigation 46: Surface Area of Power Solids

1. **(a)** Draw a net for each of the power solids that have prism-type shapes.

   (b) Discuss common features in these nets.

   (c) Using the shortest length of all power solids as one unit, calculate the area of these nets.

2. **(a)** Draw a net for each of the power solids that have pyramid-type shapes.

   (b) Discuss common features in these nets.

   (c) Using the shortest length of all power solids as one unit, calculate the area of these nets.

Investigation 47: Surface Area

1. Using power solids as a template, draw nets for: cube, square prism, triangular prism, cylinder. Use the smallest length of the solids as one unit, calculate the surface area for each of the above solids.

2. **(a)** Using power solids as a template, draw a net for a cone.

   (b) Using the method used in developing the area of a circle, find the surface area of the cone.

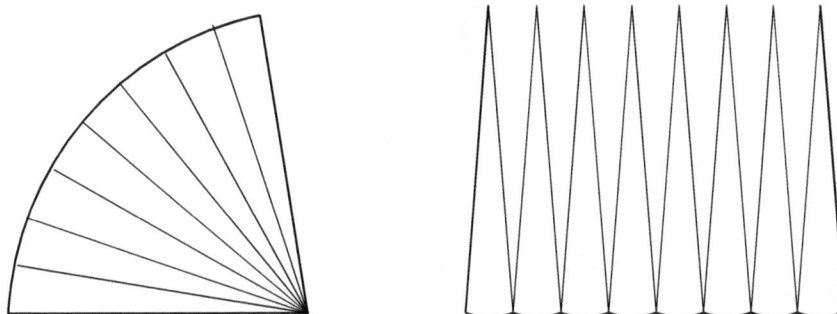

   (c) Explain how the following method can be used to find the surface are of a cone.

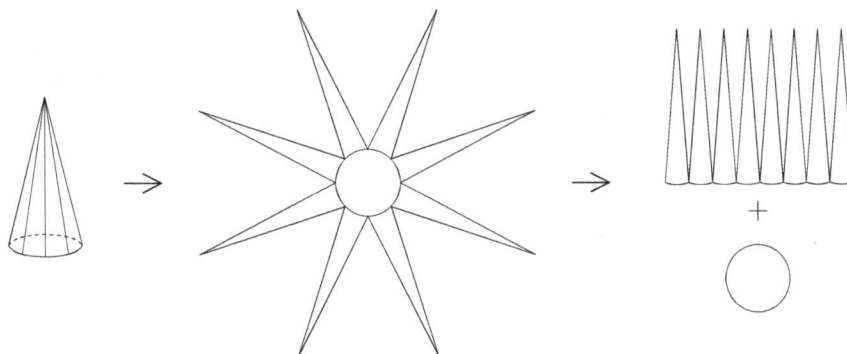

5.5 PROBLEM SOLVING

Polya's Corner

Four semi-circles are drawn with centers at the midpoints of the sides of a 2 *in* by 2 *in* square. What is the area of the shaded region?

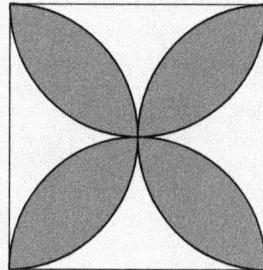

Model Polya's 4 steps:

1. Understand the problem. Can you describe the way the shaded region is formed? What are your expectations/predictions?

2. Devise a plan. What strategies could you use to solve this problem? Could you find regions on the picture that have the same area?

3. Carry out your plan.

4. Look back. Were your expectations/predictions met? Is there a simpler way to solve this problem? Can you interpret your results as a difference of the areas of two objects?

EXERCISES

1. Find the area of a rectangle with perimeter 10 *in* and with one side of length 2 *in*.

2. Find the area of the following parallelograms:

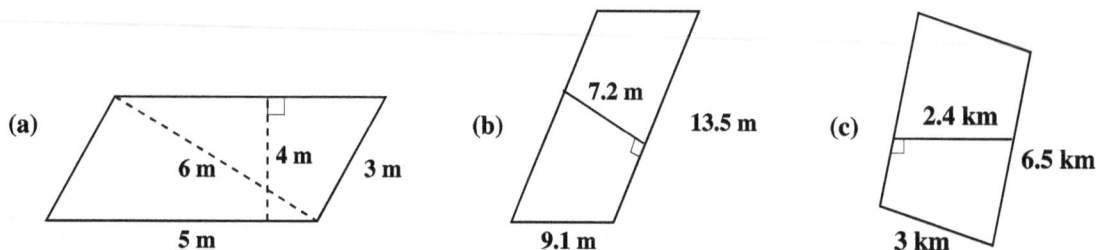

3. Find the area of the following figures:

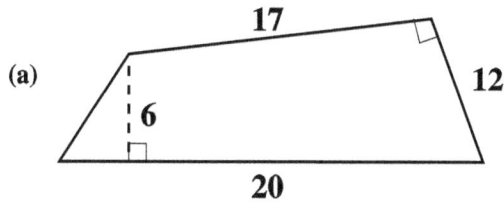

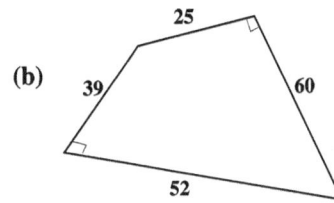

(a)

(b)

4. Explain how to obtain/deduce the formulas for the area of a trapezoid. You can assume you know all lengths you want/need.

5. Explain how to obtain/deduce the formulas for the area of a kite. You can assume you know all lengths you want/need.

6. Estimate the number π as follows:
   (a) Consider a circle of radius 1. Its circumference is 2π.
   (b) Inscribe a regular hexagon into the circle. Find its perimeter. (Hint: divide the hexagon into 6 equilateral triangles.) Let's denote this perimeter p_1.
   (c) Circumscribe a regular hexagon. Find its perimeter. (Hint: divide the hexagon into 6 equilateral triangles. Use the Pythagorean Theorem to find the length of the sides of these triangles.) Let's denote this perimeter p_2.
   (d) Notice that $p_1 < 2\pi < p_2$. What inequality do you obtain for π?

   Remark: A similar procedure can be done with regular polygons with more than 6 sides. The more sides, the harder and longer calculations, but the better the estimate. As many groups indicated in their projects, Archimedes used polygons with up to 96 sides.

7. State the Pythagorean theorem without using a diagram.

8. Write at least 3 different word problems where the solution involves the use of the Pythagorean theorem.

9. Suppose someone tells you that she has a triangle with sides having lengths $5.2, 9.5$, and 6.8. Is this triangle acute, right, or obtuse? Justify your answer.

10. A room is 36 feet long, 20 feet wide and 12 feet high. What is the distance between the two furthest corners of the room? Round your answer to the nearest foot.

11. One leg of a right triangle is 2 *cm* longer than the other leg, and the hypotenuse is 3 *cm* longer than the shorter leg. Find all sides of the triangle.

12. A cable is tied to the top of a flagpole and it is connected to the ground 6 yards away from the base of the flagpole. If the flagpole is 8 yards high, how long is the cable?

13. Find the height of a regular tetrahedron with edge 1 *cm*. Find its volume.

14. An equilateral triangle with sides 2012 units long is divided into smaller equilateral triangles with sides 4 units long (see figure in next page).

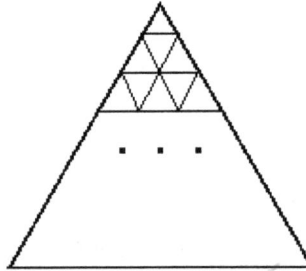

(a) How many of these small triangles are there in the big one?
(b) How many times longer is the perimeter of the big triangle than the perimeter of each small triangle in the previous problem?

15. Find the area of the following triangles.

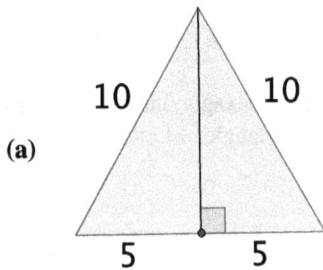

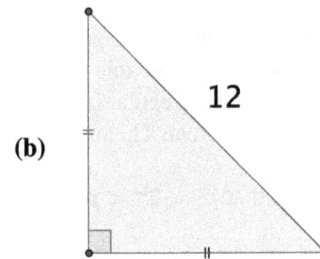

16. Find the area of the following shaded regions.

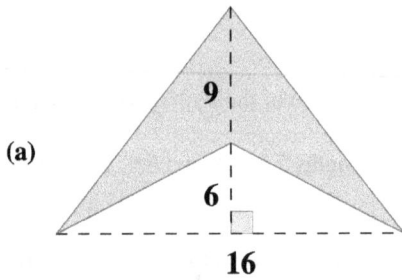

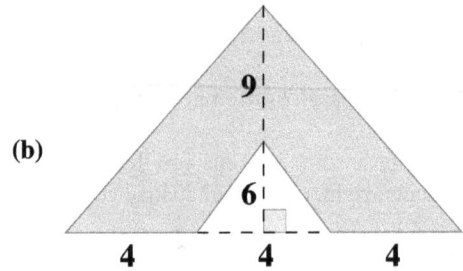

17. Find the area of the following shaded regions using the data provided:
    (a) The circle has radius 3 *units*.

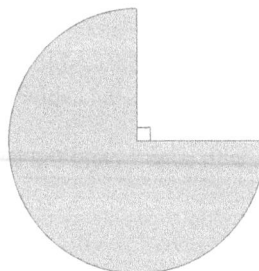

(b) Two semicircular arcs, of radius 3 *in* and 5 *in* are centered on $\overline{AB}$, which is the diameter of a larger semicircular arc (see figure below), touch tangentially at *C*.

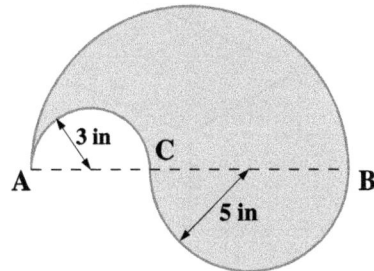

(c) The shaded region is formed by circular arcs drawn from two opposite corners of a 1 *unit* × 1 *unit* square.

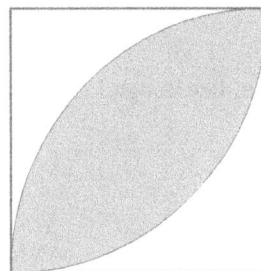

(d) The radius of the largest circle is 8 *units*. The circles are tangent to each other.

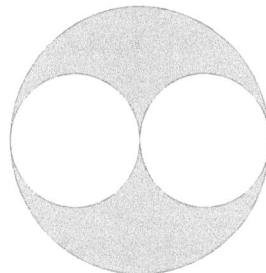

(e) The radius of the largest circle is 12 *units*. The circles are tangent to each other.

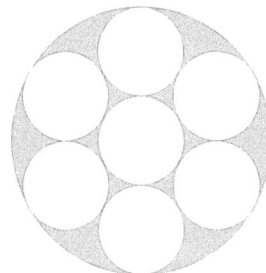

18. Consider the following picture of a house.

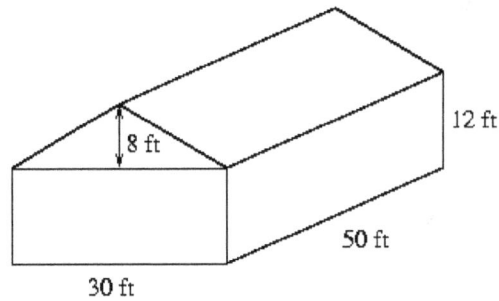

Find:
(a) Its volume.
(b) Its surface area.

19. Consider the following picture of a prism-like trough.

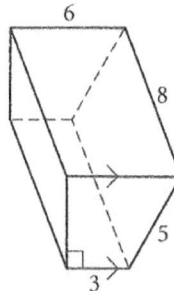

Find:
(a) Its volume.
(b) Its surface area.

20. A rectangular swimming pool with dimensions $10\ m \times 25\ m$ is being built. The pool has a shallow end that is uniform in depth and a deep end that drops off as shown in the following figure. What is the volume of this pool in cubic meters?

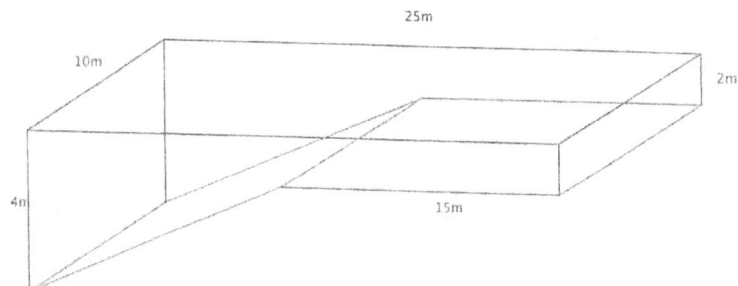

21. Find the volume of the given solid in cubic feet.

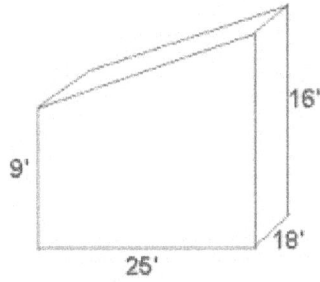

22. Find the volume of the ice cream cone pictured. Use that the top part (ice cream) is a semi-sphere.

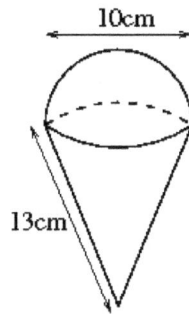

23. For the following square-based pyramid, given that $b = 66 \ ft$ and $a = 110 \ ft$.
    (a) find its volume, and
    (b) find its surface area (including the bottom.)

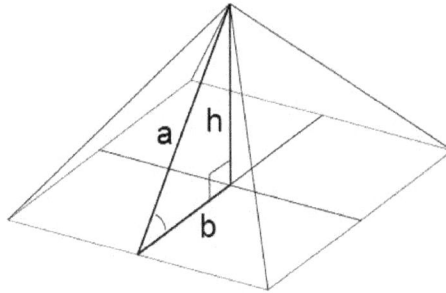

24. Determine the volume of silver needed to make the napkin ring in the following figure of solid silver. Give your answer in cubic millimeters.

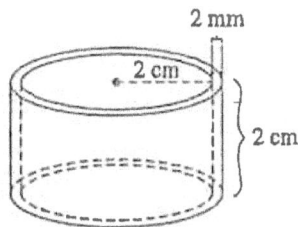

25. The following object is a rectangular prism with a cylindrical hole.

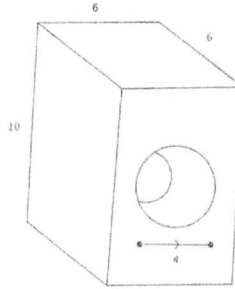

Find:
(a) Its volume.
(b) Its surface area.

26. Which is the better buy? A grapefruit 5 *cm* in radius costing 22 cents or one 6 *cm* in radius that costs 31 cents?

27. A water tank is in the shape an inverted right circular cone with diameter 10 *m* and altitude 50 *m*. Another tank of the same type is to be constructed to hold only 1/125 of the water that the old one does. If the altitude of the new tank is to be 10 *m*, what will be the diameter of the new tank? Round answer to two decimal places.

28. How does the volume compare between a triangular prism with side lengths *s* and a hexagonal prism with side lengths *s*?

29. Compare the volume of the triangular prism with side lengths *s* to the volume of a cube with side lengths *s*.

30. Compare the volume of the triangular prism with side lengths *s* to the volume of a triangular prism with side lengths 4*s*.

31. A polyhedron has volume 23 *in*^{3} and surface area 10 *in*^{2}. **Explain** how the volume and surface area are affected by quadrupling (zooming 4*x*) the shape.

32. A dog made out of Lego blocks has volume 1000 *in*^{3} and surface area 1800 *in*^{2}. **Explain** how the volume and surface area area of the dog is affected (give numbers!) by shrinking the distance between any two points to half of what it was.

33. Draw 2 different nets for a regular tetrahedron.

34. Draw 2 different nets for a square based prism.

35. Draw 2 different nets for a square based pyramid.

36. Draw 2 different nets for a rectangular solid.

37. Draw the floor view, left view, and front view of the house you live now or lived before.

38. **Challenge:** Try to make a 3 dimensional object that has a circle as floor view, triangle as right view, and rectangle as front view. Build a model (for example, carve it out of cheese, make a net and fold it from paper) of your object and bring it to class.

39. Compare the surface area of the triangular prism with side lengths s to the volume of a triangular prism with side lengths $4s$.

40. Draw the net of the following objects: rectangular prism, hexagonal prism, rectangular pyramid, hexagonal pyramid, cone, cylinder.

CHAPTER 5 REFLECTIONS

1. Explain why the area of a rectangle with lengths ℓ and width w is always equal to $\ell \times w$.

2. Using a diagram show that the area of a rectangle with length $2\frac{1}{2}$ units and width $3\frac{1}{2}$ units is equal to $2\frac{1}{2}$ *units* $\times 3\frac{1}{2}$ *units*, which is $8\frac{3}{4}$ *unit*^{2}.

3. What happens to the area of a parallelogram if it is stretched along the two parallel lines as shown below.

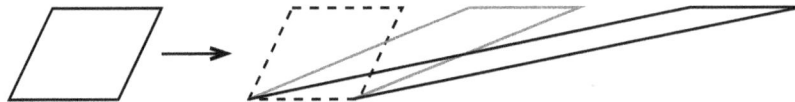

4. Use at least 3 different ways to explain why the area of the following trapezoid is $\frac{(a+b)h}{2}$.

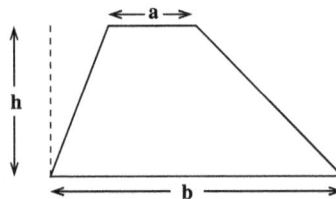

5. Using the ratio of the circumference and diameter of any circle, find the general formula for the circumference of a circle. Can you write this in terms of the radius of the circle?

6. Using dissections explain why is the area of a circle with radius 3 *in* is 9π *in*^{2}.

7. Explain why you can find the volume of a rectangular prism by multiplying the area of the base of the prism by the height of the prism.

8. We found that the volume of a rectangular prism can be calculated by multiplying the area of the base and its height. Is this true for any three dimensional object? How about a cylinder, a cone, or a pyramid?

9. How can you convince someone that an oblique prism and a corresponding right prism with same base and height have the same volume?

10. Explain the relationship of the volumes of any given prism and its corresponding pyramid (same base and height). How can this help you to find the volume of any pyramid and cone?

11. How can the concept of volume learned in this chapter be used to find the volume of standard geometric objects (i.e., prisms, cones, cylinders, pyramids, and spheres) without having to memorize formulas? Discuss how this concept can help you as a teacher.

Chapter N
Exploring Resources

N.1 SO, WHY A CHAPTER N?

This book, by no means, is meant to be a comprehensive treatment of elementary mathematics. The additional investigations, projects and resources contained in this chapter offer a wide variety of opportunities which extend inquiry into elementary level mathematics, and are intended to be utilized throughout the course. In keeping with the theme of *learning mathematics through multiple representations*, the individual and group projects, uses of technology, and material on rubrics and assessments found here aim to help students better evaluate mathematical thinking amongst their peers, thus gaining valuable experience needed for teaching their own students one day.

It is our experience that great teachers, as well as great students, are resourceful independent learners able to judge and make sense of information. Since qualities we want to cultivate in our students are qualities their teachers should also exhibit, this chapter is aptly called 'Chapter N' to reflect the continual need for teachers to seek out information and take up challenges to grow as lifelong learners in an information age, just as they intend their students to do. In this respect, although the authors cannot prepare you for everything, we have strived to provide essential skills, tools, models and methods to help prospective teachers become flexible and effective learners, as well as good communicators. Remember that teaching can be a never ending and joyful intellectual growth experience to be shared with your students; for, as the Chinese proverb goes, "to give a person a fish is to feed them for a day... but to teach them to fish is to feed them for a lifetime" (or feed them at least $N \geq 2$ days!).

N.2 FROM NUMBER BASES TO BASE-X MODEL.

After working in so many bases in chapter 3, we can now pose the following question: if we do not know what the particular base is, such as 10, 5, or 3, then can we not just call it base-X? Since in base-X, the grouping number is unknown, some modifications are necessary to representing polynomials as base-X numerals in the base bins. For example, the polynomial $2x^3 - x^2 + 1$ is represented as the base-X number $2(-1)01_X$:

x^3	x^2	x^1	$x^0 = 1$
2	-1	0	1

$2x^3 - x^2 + 1$ represented in base-X bins

To perform a calculation such as $(2x^3 - x^2 + 1) + (x^2 + 11x + 13)$ one simply converts the polynomials to their digit representations, and then employs the familiar algorithm for addition, although in the case of base-X there is no carrying or borrowing, as one does not know what the base is, so the convention is employed that within a particular bin or digit, one uses base-10 as an arbitrary frame of reference. For example,

$$(2x^3 - x^2 + 1) + (x^2 + 11x + 13) = 2(-1)01_X + 1(11)(13)_X = 20(11)(14)_X,$$

or upon putting the digits into the base-X bins, $2x^3 + 11x + 14$, as seen in the following:

x^3	x^2	x^1	x^0 $= 1$
2	-1	0	1
	1	11	13
2	0	11	14

Adding $(2x^3 - x^2 + 1) + (x^2 + 11x + 13)$ in base-X.

Working similarly to the base 5, base-X can be used to perform multiplication and division by first converting polynomials to base-X numbers, then performing the familiar algorithms used in ordinary arithmetic and converting back from base-X numbers to their algebraic polynomial representations, as in the following examples:

Example

x^5	x^4	x^3	x^2	x^1	$x^0 = 1$
			1	-11	13
		×	1	0	-1
			-1	11	-13
		0	0	0	
+	1	-11	13		
	1	-11	12	11	-13

$\left(x^2 - 1\right) \cdot \left(x^2 - 11x + 13\right) = x^4 - 11x^3 + 12x^2 + 11x - 13$ in base-X

Example

$\frac{2x^4 - x^3 + 2x^2 + x - 1}{2x - 1} = x^3 + x + 1$ in base-X long division.

N.3 PROJECTS FOR INDIVIDUAL EXPLORATIONS

1. **GeoGebra.** GeoGebra is a computer program which is an open source, easy-to-use, versatile tool for visualizing mathematical concepts from elementary through college level. GeoGebra offers the opportunity for teachers to create interactive online learning environments to foster experimental and discovery learning. In addition, GeoGebraWiki and the user forum provide platforms for sharing free interactive teaching materials and obtaining support from fellow users.

   (a) Go to the website of GeoGebra at `http://www.geogebra.org/cms` and download the program.
   (b) Experiment how various tools work. Learn about developing dynamic worksheets by clicking on *help*, and then studying *introduction to GeoGebra*. Complete at least 4 activities from *Practice block I* and *Practice block II*. Print out the results of your activities.
   (c) Go to the website of GeoGebra: `http://www.geogebra.org/cms`. You will see at the top right side a *WIKI* button to go to the GeoGebra Wiki. There are plenty of teaching materials (dynamic worksheets) uploaded there by various users on various languages. Find at least two dynamic worksheets you believe you could use well in your instruction of mathematics. Print out the worksheets, and write 2-3 paragraphs about your plans of using each of them.

2. **Metric System.** Answer in writing the following questions after studying the information provided about the International System of Units on the web. For example, the following website might be useful:
   `http://en.wikipedia.org/wiki/International~System_of_Units`

   (a) Why and by whom was the metric system developed?
   (b) What is the relationship between the metric system and the International System of Units?
   (c) What is more? 5 hecto Pascal or 5 Newton per cm^2? Justify your answer.
   (d) Express the speed of 65 miles per hour in the international system of units.
   (e) List three facts you learned from the reading.

3. **Surface Area.** Make at least three different open-top containers using a sheet of 8.5 inch by 11 inch paper for each container. You can use Scotch tape to stabilize your container, but the outside surface of your container should be all paper.

   (a) Draw a net for each of your containers.
   (b) Calculate the surface area of each of your containers.
   (c) Calculate the volume of rice each of your containers could hold.
   (d) Design an open-top container that you could build from a sheet of 8.5 inch by 11 inch paper that you believe would hold the largest volume of rice. Measure in liters the amount of rice you can pour into it without 'overflow.' Sketch a picture of your container.

4. **Web Search.**

   (a) Find at least 3 websites that have valuable information for you about the use of manipulatives and/or technology in mathematics instruction. Write a 2-3 pages long paper about your ideas of using the information you found.
   (b) Find at least 3 websites that have valuable information for you about questioning techniques in mathematics instruction. Write a 1-2 page paper comparing your questioning practices with the promoted practices.
   (c) Learn about various competitions for elementary school students on the web. One great competition is called *Abacus*, that you can find on the following website:
   `http://www.gcschool.org/program/abacus/index.aspx`

Read about the Abacus competition, and check out some of the archive and current problems. Solve at least 4 problems from the archive problems and 4 problems from the most current problems. Make sure to use sound reasoning, and try to find alternative solutions, too!

5. **Data Analysis.** A course is taken by 4 students: Alice, Bruce, Chloe, and Daniel. There are two tests, homework, and a final exam given to students. The first test is 20% of the final grade, the second test is 30% of the final grade, the homework is 15% of the final grade, and the final exam is 35% of the final grade.

The instructor gives
A for 87% − 100%
B for 75% − 86%
C for 62% − 74%
D for 50% − 61%
F for 0% − 49%

Alice got 38 points out of 50 points on the first test, 66 points out of 100 points on the second test, 28 points out of 40 points on the homework, and 167 points out of 200 points on the final exam.

Bruce got 42 points out of 50 points on the first test, 70 points out of 100 points on the second test, 34 points out of 40 points on the homework, and 187 points out of 200 points on the final exam.

Daniel got 41 points out of 50 points on the first test, 92 points out of 100 points on the second test, 37 points out of 40 points on the homework, and has not taken the final exam yet.

(a) Calculate the final grades of Alice and Bruce.
(b) How many points does Daniel need to get on the final exam in order to get an A on the course?
(c) How many points does Daniel need to get on the final exam in order to get a B on the course?
(d) The mean of the scores on the first test was 39. Could you figure out Chloe's score on the first test? If yes, explain how, if no, explain why not?
(e) The median on the second test was 77 points. Could you figure out Chloe's score on the second test? If yes, explain how, if no, explain why not?
(f) The median on the homework was 31 points. Could you figure out Chloe's score on the homework? If yes, explain how, if no, explain why not?

6. **Polygons.**

(a) It is easy to see that the sum of the interior angles of a regular quadrilateral (square) is $360°$. Find the sum of the interior angles of a regular n-gon (i.e. a regular polygon with n vertices).
(b) Note that the sum of the interior angles of a triangle does not depend on its shape. It does not matter whether the triangle is regular, acute, or obtuse – the angle sum is always the same, $180°$. Given a whole number $n > 3$, would the sum of the interior angles be the same for any n-gon, or will it depend on whether the n-gon is regular or irregular, convex or concave? Justify your answer!
(c) Do your strategies/solutions work for any n-gon? For the first part, what can you say about your answer (formula)? Does it make sense? Were your expectations met? The two parts seem to be somewhat related. Did you use same or different ideas to solve them? If different, can you adapt each idea for the other part? Why or why not?

For this problem you need to recall that the sum of the interior angles of any triangle is $180°$.

N.4 GROUP PROJECTS

Project 1: Numbers

Use the Internet to find factual information about the numbers (Topic) assigned to your group. Turn in an approximately 3-page paper on the results of your investigation (double-spaced, 12 pt.). A few questions are given to guide you, but feel free to touch other topics. The paper should be written by the group. Printouts or reproductions of existing sites are not acceptable, however, you can quote different sources, or use pictures/diagrams (as long as you name the source) to illustrate your paper. Name at least 3 different sources (web sites) that you have used to collect the information. Please remember to put the names of all members of your group on your paper. Finally, prepare a short (about 5-8 minutes) presentation using a poster and any additional materials you feel appropriate (e.g. power point, manipulatives, models, handouts, etc.).

Topic A: Complex Numbers

- Give the definition and examples.
- What notation is used in mathematics for the set of all complex numbers?
- Which culture/individual(s) were credited with discovering/using complex numbers first?
- Why were complex numbers needed?
- Name various properties of complex numbers.
- Explain how to add two complex numbers and how to multiply two complex numbers.

Topic B: Prime Numbers

- Give precise definition and examples.
- Which culture/individual(s) were credited with contributing to the knowledge about prime numbers?
- What is the Sieve of Eratosthenes?
- When and how was it proven that there are infinitely many prime numbers? (Give a proof in your paper.)

Topic C: $\sqrt{2}$

- Give the definition and an approximate value of $\sqrt{2}$.
- Which culture/individual(s) were credited with discovering/using $\sqrt{2}$ first?
- When and how was it proven that $\sqrt{2}$ is irrational? (Give a proof in your paper.)
- Why is $\sqrt{2}$ important in mathematics?
- Name some properties of $\sqrt{2}$.

Topic D: The number π

- What is an approximate value of π?
- Which culture/individual(s) were credited with discovering/using π first?
- Who and when first proved that π is irrational?
- Name various properties of π.

Project 2: The Pythagorean Theorem

Your group is assigned to explore and learn about the Pythagorean theorem using the software called *GeoGebra*.

You may get access to the program from `www.geogebra.org`. Get familiar with the various functions of the program. This program will be a very useful tool for you as a teacher in the future.

PART A

1. Construct a right triangle (make sure you can make the triangle big or small by dragging a blue color point while keeping the right angle). You may use the *Perpendicular Line* tool to make a right triangle.

2. Using the *Regular Polygon* tool, mount squares on each side of the right triangle.

3. Using the *Area* tool, measure the area of each square.

4. Using the *Input* box at the bottom of the page, add the areas of the two small squares to show that this sum is equal to the area of the larger square.

5. Use the *Insert Text* tool to write down your name and an explanation of what this diagram tells you in relation to the Pythagorean theorem.

PART B

1. Construct another sketch to explore if what you showed in part A is true even if any other shape (pentagon, triangle, hexagon, semi circle) is mounted on the right triangle instead of a square.

2. Explain your reasoning for your answer to part 1 using mathematical calculations and logical thinking.

Along with the printouts of your sketches, turn in a 1-page reflection explaining how you can use what you have learned in this investigation when teaching in an elementary school mathematics classroom. Make sure to address how *GeoGebra* can be useful for you as a teacher.

N.5 MULTIPLE REPRESENTATIONS

Explain how these four representations are related to each other.

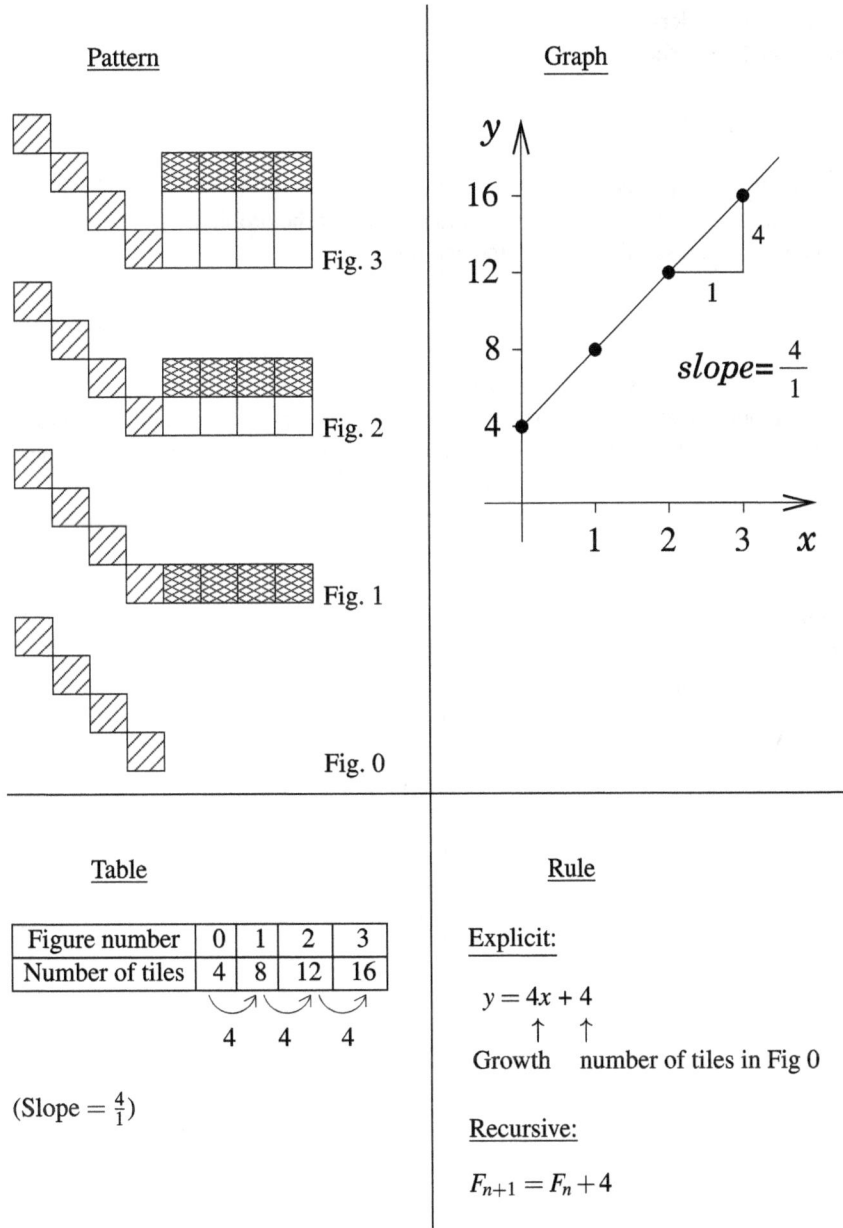

Pattern

Fig. 3

Fig. 2

Fig. 1

Fig. 0

Graph

$$slope = \frac{4}{1}$$

Table

Figure number	0	1	2	3
Number of tiles	4	8	12	16

4   4   4

(Slope $= \frac{4}{1}$)

Rule

Explicit:

$$y = 4x + 4$$

↑   ↑
Growth   number of tiles in Fig 0

Recursive:

$$F_{n+1} = F_n + 4$$

N.6 RUBRIC FOR GRADING PROBLEM SOLVING TASKS

(Adapted from the work of Van de Valle, 2008)

UNDERSTANDING THE PROBLEM

0 Complete misunderstanding of the problem
1 Part of the problem misunderstood or misinterpreted
2 Complete understanding of the problem

PLANNING THE SOLUTION

0 No attempt, or totally inappropriate plan
1 Partially correct plan, based on correct interpretation of part of the problem
2 Plan could lead to a correct solution if implemented properly

GETTING AN ANSWER

0 No answer, or wrong answer based on an inappropriate plan
1 Copying error, computational error, or partial answer for a problem with multiple answers
2 Correct implementation of the plan (regardless of correctness of the plan), and correct label for the answer

LOOKING BACK

0 No evidence of analysing the results
1 Partial, but incomplete analysis
2 Complete analysis of the answer as well as the solution process

N.7 ADDITIONAL CHALLENGES

Problems from competitions for 6-graders

1. The figure shown consists of 8 congruent squares. The perimeter of the figure is 36 units. What is the area of the figure?

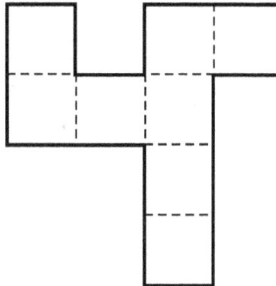

2. The cafeteria sells each apple at one price and each banana at another price. For 1 apple and 3 bananas Jose pays $2.05. For 4 apples and 2 bananas April pays $2.70. Maria buys 2 apples and 2 bananas. How much does she have to pay?
3. There are 8-legged spiders and 6-legged flies in a room. The total number of insects is 80. The total number of insect legs is 604. How many more spiders than flies are in the room?
4. A tractor has 14 gallon gasoline tank. The tractor starts with a full tank of gasoline. It runs out of gasoline when it is done plowing 3/5 of a field. How much gallons of gasoline does the tractor need to plow the whole field?
5. If the length of a rectangle is reduced by 50% and the width of the rectangle is increased by 50%, how does the area of the rectangle change?
6. An animal walks 30 feet in 5 seconds. What is the speed of this animal in miles per hour?

Proportional thinking

1. Solve the following problem in two ways. One of the ways must be using proportions. You should address if the problem is direct or inverse proportion? Explain. Then set a proportion and solve it to get an answer. A Fresno State student notices that for every 5 hours a week at his/her job his/her GPA goes down 1.3 points. This semester he/she will work 23 hours a week. What will be his/her GPA at the end of the semester if at the beginning it was a 3.5?

2. Solve the following problem in two ways. One of the ways must be using proportions. You should address if the problem is direct or inverse proportion? Explain. Then set a proportion and solve it to get an answer. One painter takes 3 hours to paint a wall, how much would it take 5 painters to paint the same wall?

3. Solve the following problem in two ways. One of the ways must be using proportions. You should address if the problem is direct or inverse proportion? Explain. Then set a proportion and solve it to get an answer. Jim found out that after working for 9 months he had earned 6 days of vacation time. How many days will he have earned after working for two years?

4. Write a story problem that is solved by using the following proportion. Explain clearly what quantities are related in the proportion, and what type of proportion this problem uses (direct or inverse).

$$\frac{3}{5} = \frac{15}{x}$$

5. Write a story problem that is solved by using the following proportion. Explain clearly what quantities are related in the proportion, and what type of proportion this problem uses (direct or inverse).

$$\frac{12}{25} = \frac{x}{13}$$

6. Write a story problem that is solved by using the following proportion. Explain clearly what quantities are related in the proportion, and what type of proportion this problem uses (direct or inverse).

$$\frac{31}{5} = \frac{17}{x}$$

7. What is the 75% of 15? 3 is what percent of 14? 75% of what number is 37?

8. In the last year Carmen has driven $11,645$ miles and she has changed the oil four times. If she continues maintaining her car at this rate, and she only drives $5,700$ miles next year, about how many times will she change the oil?

9. Imagine that the teachers in your school decide to play the lottery together. If they win, the prize is $800,000. How much money should each get if there are 2 teachers? 5 teachers? 17 teachers?

10. Solve the following problem in two ways.
    20 men produce 3000 articles in 12 days. How long should 15 men take?

11. A piece of computer software is to be developed by a team of programmers. It is estimated that a team of four people would take a year. How long would it take three programmers to do the same job?

12. A 10 pounds bag of potatoes lasts for a week when used in catering for 7 people.
    (a) How long will it last for 2 people, assuming everybody eats the same amount?
    (b) If, instead of buying a 10 pounds bag (which might not keep well), you want to buy fresh potatoes every week, how much per week should you buy for 2 people?

13. Two workers in Fresno State's warehouse take 20 minutes to stick labels on 500 packages. There are still 4000 more packages. How many workers are required, if the job is to be done in about a further hour?

14. A carpenter has three large boxes. Inside each box are two medium-sized boxes. Inside each medium-sized box are five small boxes. How many boxes are there altogether?

15. Suppose you are given an unfamiliar book. You have 10 minutes to estimate (not just guess!) how long it will take you to read this book. How can you do this? (Hint: use proportional reasoning!)

16. Solve the following problem in at least two different ways. At least one of your solutions should be very simple (accessible to a first-grader, i.e. should not use fractions, etc).
    Aby can run 5 laps in 12 minutes. Ben can run 6 laps in 14 minutes. Who is the faster runner?

17. Two camps of Scouts are having pizza parties. The Bunny Camp ordered enough so that every 5 campers will have 2 pizzas. The Fox Camp ordered enough so that every 7 campers will have 3 pizzas. If within each camp the pizzas are split equally, campers of which camp will eat more pizza?

18. A mule and a horse were carrying some bales of cloth. The mule said to the horse, "If you give me one of your bales, I shall carry as many as you." "If you give me one of yours," replied the horse, "I will be carrying twice as many as you." How many bales was each animal carrying? Find as many different solutions as you can.

19. Write the following polynomials using the base-X model:

   (a) $x^3 + x + 1$ **(b)** $x^5 - x^3 + x^4 + 3$ **(c)** $13 + 2x^4 - 21x + 3x^2$

20. Perform the following operations with polynomials using with the base-X model:

   (a) $(x^5 - x^3 + x^4 + 3) + (2x + 2x^3 + x^2 + 3x^5)$ **(b)** $(x^2 - 2x^3 + 4) - (3x - x + 1 - 2x^2)$

   (c) $(1 + 2x + 3x^2)(2x^4 + x^2 + 2x + 2)$ **(d)** $(2x + 1)^4$

   (e) $\dfrac{x^2 + 5x + 6}{x + 2}$ **(f)** $\dfrac{2x^2 + 3x + 1}{x + 1}$

   (g) $\dfrac{2x^3 + 3x^2 + 4x - 3}{2x - 1}$

N.8 ADDITIONAL CHALLENGES WITH SOLUTIONS

1. **Mathematicians at the car wash:** Abel, Bolzano, Cauchy, Dedekind and Euclid were washing cars to gather some money to buy math books.
   Abel washed half of the cars that were left at the car wash, then Bolzano washed a third of the ones that were left by Abel, after him Cauchy washed just a fourth of what Bolzano had left unwashed. Finally Dedekind and Euclid washed the last three cars. How many cars were left at the car wash?

2. **Counterfeit coin I:** In a group of five 'gold' coins, one of them is counterfeit, and thus weights more than the rest but looks exactly like the others. Using a pan balance, what is the smallest number of balancings needed to identify the fake gold coin?

3. **Numerical letters:** Assume that each letter F, I, M, N, S, U, W represents a digit from $1, 2, 3, 6, 7, 9, 0$. Find the value of each letter if you know that

$$
\begin{array}{r}
S\ U\ N \\
+\ F\ U\ N \\
\hline
S\ W\ I\ M
\end{array}
$$

4. **The four 4's I:** Write each integer from 0 through 25 using exactly four number 4's. You can use any matematical operations $(+, -, \times, \div, sqrt, !,$ exponents$)$ and parentheses.
   For example $0 = (4 - 4) + (4 - 4) = (4 \div 4) - (4 \div 4)$.

5. **Factorizations of a number:** In how many ways can you factor (using only natural numbers) the number 1000.

   Too hard? Try an easier problem first!

   In how many ways can you factor 24?

6. **Magic triangle:** Place the numbers $1, 2, 3, 4, 5, 6, 7, 8, 9$ in the circles below so that the sum of the numbers in each side of the triangle is 17.

   Too hard? Try an easier problem first!

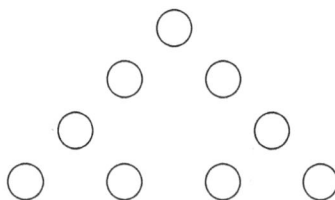

Place numbers $1, 2, 3, 4, 5, 6$ in the circles of the following triangle so the sum in each side is 12.

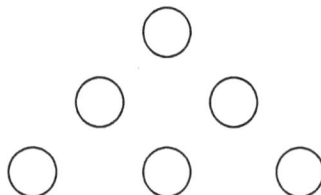

7. **Sum off odd numbers:** Look at the picture below. Can you guess what the sum of the first many odd numbers should be?

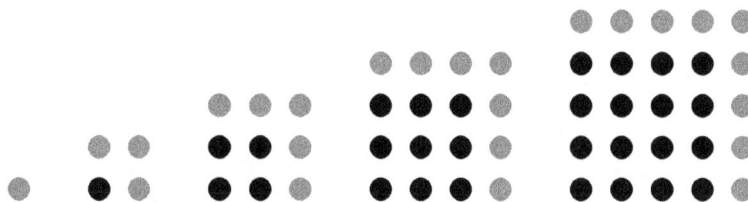

Too hard? Try smaller simpler cases first!

8. **How many? How many?:** How many angles are there in the figure?

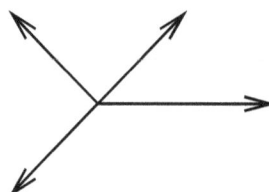

   (a) 4
   (b) 6
   (c) 7
   (d) 8
   (e) More than 8

9. **The end of squares:** Is it possible for a square number to end in a 2? In a 3? What are the digits allowed to be at the end of a square number?

10. **Pizza for eleven:** Can we cut a pizza into exactly 11 pieces (they do not have to be of the same size) with exactly 4 straight cuts?

11. **Sum of four consecutive numbers:** Is the sum of four consecutive whole numbers always even? Why?

    Too hard? Try a baby intermediate case!

    Is the sum of two consecutive odd numbers always even?

12. **Darts:** In a darts game, three darts are thrown. All hit the target (see picture below). What scores are possible?

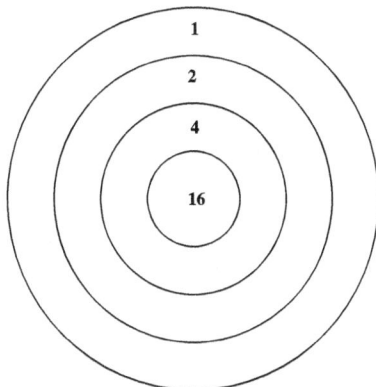

13. **Fermat's pet store:** When Fermat opened his pet store he had 128 cats, after a week he had 64, the week after he had only 32. If this keeps going in how many weeks he would have ran out of cats?

14. **Counterfeit coin II:** In a group of nine 'gold' coins, one of them is counterfeit, and thus weights more than the rest but looks exactly like the others. Using a pan balance, what is the smallest number of balancings needed to identify the fake gold coin?

15. **+ and - with pattern blocks:** Assume that two yellow (pattern block) hexagons form a whole. What are the other figures?

Green triangle: _____   **Red trapezoid:** _____
Blue parallelogram: _____   **Yellow hexagon:** _____

Now use these figures to compute and explain the operations.

(a) $\dfrac{5}{6} + \dfrac{1}{2}$ (b) $1\dfrac{1}{6} - \dfrac{3}{4}$ (c) $1\dfrac{5}{6} - \dfrac{3}{2}$

16. **Geoboard I:** Find the areas of the figures given below. Then find general formulas for such figures (kite, rhombus, trapezoid, triangle).

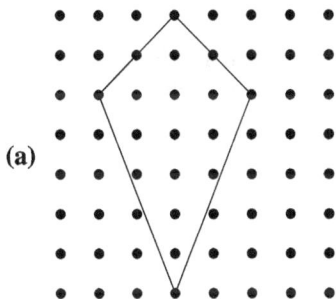

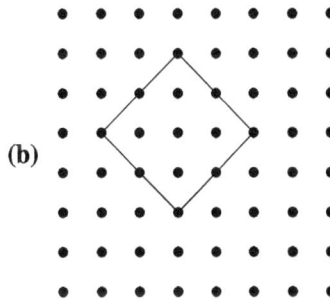

(c)

(d)

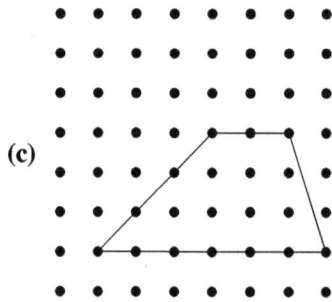

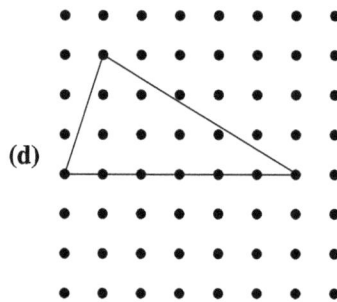

17. **Geoboard II:** What are the areas, and perimeters, of the figures below?

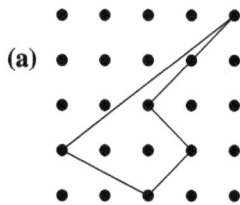

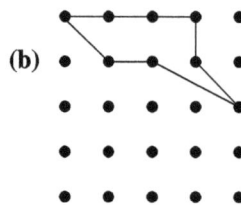

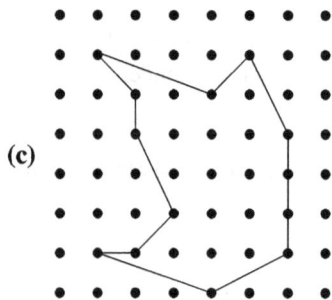

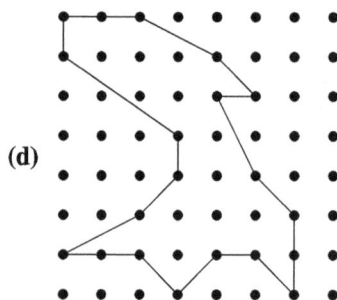

(a)

(b)

(c)

(d)

18. **Sewing:** Sonya K. bought a piece of cloth 48 inches wide and 1 yard long. It cost $12. She cut off one-fourth and used it to make a tablecloth. From the remaining material, she used a piece that was 12 inches wide and 12 inches long to make a scarf and a piece 1.5 feet by 2 feet to make the cover for a pillow. Her friend Emmy N. saw Sonya's sewing efforts and said she really liked the material. She wanted to buy what was left to do some sewing of her own. Sonya was willing to sell leftover material for the rate that she had paid for it. How much should she charge Emmy?

19. **Balancing books:** A balance scale was in perfect balance when Gauss placed a math book on one pan of the balance and 3/4 of the same-sized book together with 3/4-pound weight on the other pan. How much did the book weight?

20. **Pattern blocks I:** Use pattern blocks to represent the following fractions

$$\frac{1}{2}, \frac{1}{3}, \frac{1}{4}, \frac{1}{6}, \frac{1}{12}$$

and any other you are able to do.

21. **Shaded circle:** What part of the circle has been shaded?

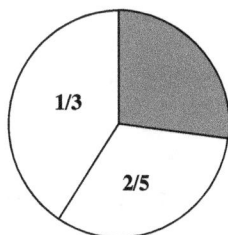

Show step-by-step the thinking you did to figure this out.

22. **Pattern blocks II:** Use pattern blocks (and any other way you want) to calculate, and explain, the following computations

   (a) $\dfrac{1}{5} + \dfrac{2}{3}$ (b) $\dfrac{2}{5} - \dfrac{1}{4}$

   In particular, you should be able to explain how to find the common denominator, which is needed to solve these problems.

   Finally, make up problems that need the calculations above to be solved.

23. **Mind reader I:** One half of the number I am thinking of is $\frac{1}{3}$. What is that number?

24. **Mind reader II:** One third of the number I am thinking of minus a half of that same number is 8. What is the number?

25. **Be proper:** Explain why $2\frac{1}{3} = \frac{7}{3}$.

26. **A picture is worth a thousand words:** How does this picture

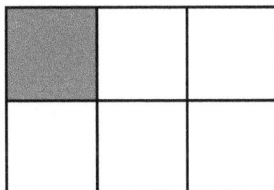

   explain that

$$\frac{1}{2} \cdot \frac{1}{3} = \frac{1}{6} \, ?$$

27. **Simple math:** What is $\frac{2}{3}$ of 12? of 14?

28. **Fighting about pizza:** Frank ate 12 pieces of pizza and Dave ate 15 pieces. "I ate $\frac{1}{4}$ more", said Dave. "I ate $\frac{1}{5}$ less", said Frank. Who was right?

29. **Wrong but right:** A math trick shows that a fraction $\frac{16}{64}$ can be simplified by 'canceling' the 6's and obtaining $\frac{1}{4}$. There are many other fractions for which this technique yields the correct answer.

   (a) Apply this technique to each of the following fractions and check that the results are correct

   (i) $\dfrac{16}{64}$ (ii) $\dfrac{19}{95}$ (iii) $\dfrac{26}{65}$ (iv) $\dfrac{199}{995}$ (v) $\dfrac{26666}{66665}$

   (b) Using the pattern established in parts **iv** and **v** of part (a), write three more examples of fractions for which this method of simplification works.

30. **Grandma's quilt:** Grandma was planning to make a red, white and blue quilt. One-third was to be red and two-fifths was to be white. If the area of the quilt was to be 30 square feet, how many square feet would be blue?

31. **In base 5:** Perform the following 3 computations in base 5. Explain any 'mysterious', or different from standard operations, step. Do NOT transform to base 10 to perform the computations, you can do this to check your answers if you please.

    (a) $122_5 + 333_5$ **(b)** $222_5 - 133_5$ **(c)** $22_5 \times 33_5$

32. **Grade conscious student:** A mathematics student notices that after every 3 days studying for a quiz his/her grade in the quiz goes up 5 points. How many points will his/her grade go up after studying for two weeks straight?
    Solve the previous problem in two ways. One of the ways must be using proportions. You should explain whether the problem is direct or inverse proportion. Then set a proportion and solve it to get an answer.

33. **Doing the laundry:** You buy a family-size box of laundry detergent that contains 40 cups. If your washing machine calls for $1\frac{1}{4}$ cups per wash load, how many loads of wash can you do?

34. **Baking cookies:** A recipe that makes 3 dozen peanut butter cookies calls for $1\frac{1}{4}$ cups of flour.

    (a) How much flour would you need if you doubled the recipe?
    (b) How much flour would you need for half the recipe?
    (c) How much flour would you need to make 5 dozen cookies?

35. **Catch the errors:** What follows are examples of student work in multiplying fractions. In each case, identify the error and answer the given problem as the student would:

    Sam:
    $$\frac{1}{2} \times \frac{2}{3} = \frac{3}{6} \times \frac{4}{6} = \frac{12}{6} = 2$$

    $$\frac{3}{4} \times \frac{1}{8} = \frac{6}{8} \times \frac{1}{8} = \frac{6}{8} = \frac{3}{4}$$

    $$\frac{3}{4} \times \frac{1}{6} = ?$$

    Tom:
    $$\frac{3}{8} \times \frac{5}{6} = \frac{3}{8} \times \frac{6}{5} = \frac{18}{40} = \frac{9}{20}$$

    $$\frac{2}{5} \times \frac{2}{3} = \frac{2}{5} \times \frac{3}{2} = \frac{6}{10} = \frac{3}{5}$$

    $$\frac{5}{6} \times \frac{3}{8} = ?$$

    Each student is confusing multiplication with another algorithm. Which one?

36. **Chickens:**
    (a) A chicken and a half lays an egg and a half in a day and a half. How many eggs do 12 chickens lay in 12 days?
    (b) How long would it take 3 chickens to lay 2 dozen eggs?
    (c) How many chickens will it take to lay 36 eggs in 6 days?

37. **Guests at a dinner:** How many guests were present at a dinner if every two guests shared a bowl of rice, every three guests shared a bowl of broth, every four guests shared a bowl of fowl, and 65 bowls were used altogether?

38. **In between:** Find three fractions that are greater than $\frac{2}{5}$ and less than $\frac{3}{7}$.

39. **Agnes's running habits:** Agnes runs on the same route every morning, starting at the same time. Her route includes running through a bridge that has railroads tracks, too. Each morning there is a passenger train crossing the bridge while Agnes is running. Agnes knows that the train whistles when she has passed 3/8 of the length of the bridge. She has experienced several times that if she keeps running with full speed forward, the train reaches her exactly at the moment she gets the further end of the bridge. On the other hand, if she turns around at the moment she hears the whistle and keeps running with full speed towards the closer end of the bridge, she meets the train exactly at the closer end of the bridge. The 'full speed' Agnes can maintain is 10 miles per hour. What is the speed of the train?

40. **Birthday cake:** Mom made a rectangular birthday cake and said I could share it with my friends. Adnan took 1/6 of the cake, Bob took 1/4 of what was left. Carmen cut 1/4 of what is remained and have Doreen before helping himself 1/3 of the remaining cake. After they had all eaten their cake, my best friend and I split the piece that was left in the pan. Did everyone get a fair share? Explain your reasoning.

41. **SpongeBob:** On dry land, SpongeBob is 20% water and weighs a total of 2 ounces. After going to the Krusty Krab (down in the ocean), he gets back on land but is now 85% water! What is SpongeBob's total weight now?

42. **Algebra tiles I:** Use algebra tiles to do what is asked

        (a) factor $2x^2 - 5x + 2$ **(b)** multiply $(x - 3)(2x + 1)$

43. **Algebra tiles II:** Solve the equation $x^2 - 5x + 4 = x - 1$ by completing the square or factoring (using algebra tiles). Then use the quadratic formula or any other method to check your answer.

44. **How old am I?:** Find two **distinct** solutions for the following problem. One of them must involve setting and solving an equation.
Fifteen years ago my mom was twice as old as I was. Now her age is just three-halves of my age. How old am I?

45. **Isosceles triangle area:** Find the area of triangle ABC where $\angle ABC = 35°$, $\angle CAB = 35°$, length $BC = 5\ in$, and length of AB is 6.

46. **Volume of a trough:** There is a trough that is 7 feet long, and its vertical cross sections are inverted isosceles triangles with a height 5 feet and base 2 feet.
What is the volume of the trough?

47. **Length of a ladder:** A ladder leans against a vertical wall. The top of the ladder is 5 yards above the ground. When the bottom of the ladder is moved 2 yards farther away from the wall, the top of the ladder rests against the foot of the wall. What is the length of the ladder?

48. **Chevy and Ford:** A Chevy starts traveling east along a road. At the same time, from the same point a Ford starts traveling north at a speed 60 mi/hr. After one hour and twenty minutes, the cars are 100 mi apart. At what speed is the Chevy traveling?

49. **Width of a river:** A surveyor places a base line along one bank of a river. From each end of the base line, a rock is sighted on the opposite bank of the river right on front of the base line. The base line is 60 yards long and the lines of sight of the rock form angles of $60°$ with the base line. How wide is the river?

50. **Irrigation problem:** In Fresno many fields have circular irrigation systems. A single sprinkler is placed on the center of the field. If the sprinkler sprays water just to touch the four edges of the field, what percentage of the field is not watered?

51. **Zooming in:** A picture that measures 12 cm by 18 cm is enlarged to 4 times its area. What are the new dimensions?

52. **Squares in squares in squares:** Each of the squares shown is inscribed in a larger square so that the vertices of the inscribed square bisect the sides of the larger square. What fraction of the area of the largest square is shaded? Express your answer in simplest form.

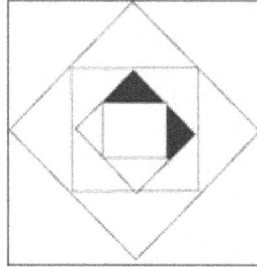

53. **Star-shaped area:** Line segments $\overline{AB}$ and $\overline{CD}$ each have length 2, are perpendicular, and bisect each other. Each of the four arcs is an arc of a circle, which is tangent to the two line segments. Find the area of the shaded region.

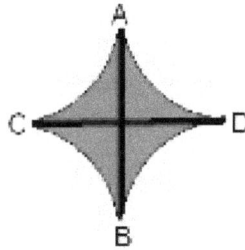

SOLUTIONS / HINTS

What follows is just the answers or **a** solution to the problems assigned (for you to check your work). In any case, don't forget that you should try to solve these problems in as many ways as possible, and also you should be able explain your solution to anybody.

1. **Mathematicians at the car wash:** Abel washed 6 cars, Bolzano washed 2, Cauchy 1 car, Dedekind and Euclid the last three. There were 12 cars left at the car wash.

2. **Counterfeit coin I:** Take four of the five coins, get two groups of four and weight. If the scale is balanced then you know that the counterfeit coin is the fifth (that stayed out). In case one of the sides of the scale is heavier, then you know that the fake coin is on that side. To determine which one is the heaviest you just need to compare these two coins on the scale...Done. So, two balancings are enough.

3. **Numerical letters:** $1 = S, 2 = M, 3 = U, 6 = N, 7 = I, 9 = F, 0 = W$. Thus,

$$\begin{array}{r} 1\,3\,6 \\ +\,9\,3\,6 \\ \hline 1\,0\,7\,2 \end{array}$$

4. **The four** 4'**s I:** Here is one (of the many) solutions

$$
\begin{aligned}
0 &= (4-4)+(4-4) \\
1 &= (4-4)+(4\div4) \\
2 &= (4\div4)+(4\div4) \\
3 &= (4+4+4)\div4 \\
4 &= 4+(4-4)\div4 \\
5 &= ((4\times4)+4)\div4 \\
6 &= 4+(4+4)\div4 \\
7 &= (4+4)-(4\div4) \\
8 &= (4+4)+(4-4) \\
9 &= (4+4)+(4\div4) \\
10 &= 4\times\sqrt{4}+4\div\sqrt{4} \\
11 &= (4!+4!-4)\div4 \\
12 &= 4\times(4-(4\div4))
\end{aligned}
\qquad
\begin{aligned}
13 &= (4!+4!+4)\div4 \\
14 &= 4+4+4+\sqrt{4} \\
15 &= (4\times4)-(4\div4) \\
16 &= 4+4+4+4 \\
17 &= (4\times4)+(4\div4) \\
18 &= 4\times4+4-\sqrt{4} \\
19 &= (4!-4)-(4\div4) \\
20 &= 4!+4-4+4 \\
21 &= (4!-4)+(4\div4) \\
22 &= 4\times4+4+\sqrt{4} \\
23 &= 4!-4^{4-4} \\
24 &= 4!+4\times(4-4) \\
25 &= 4!+((\sqrt{4}\times\sqrt{4})\div4)
\end{aligned}
$$

5. **Factorizations of a number:** You play with the factoring of 1000. For 24 we get

$$24 = 24\cdot1 = 12\cdot2 = 6\cdot4 = 3\cdot8 = 6\cdot2\cdot2 = 3\cdot4\cdot2 = 3\cdot2\cdot2\cdot2$$

6. **Magic triangle:** Here is one of the solutions to the big triangle problem

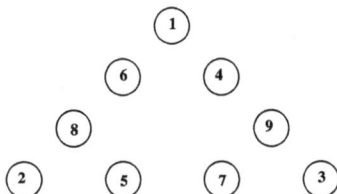

Can you find the other?

7. **Sum off odd numbers:** We will talk about this problem later on. Right now you should realize that the sum of

    the first 2 odd numbers is $4 = 2^2$

    the first 3 odd numbers is $9 = 3^2$

    the first 4 odd numbers is $16 = 4^2$

    There seems to be a pattern here... how does that relate to the picture?

8. **How many? How many?:** More than 8

9. **The end of squares:** If one makes a list of the first many squares one gets

$$1, 4, 9, 16, 25, 36, 49, 64, 81, 100, 121, 144, 169, 196...$$

   we see that the final digits are always $0, 1, 4, 5, 6$ and 9. Is this always true? We look at the numbers and realize that

   (a) numbers ending in a 0 square to a number ending with a 0,

   (b) numbers ending in a 1 square to a number ending with a 1,

   (c) numbers ending in a 2 square to a number ending with a 4,

   (d) numbers ending in a 3 square to a number ending with a 9,

   (e) numbers ending in a 4 square to a number ending with a 6,

   (f) numbers ending in a 5 square to a number ending with a 5,

   (g) numbers ending in a 6 square to a number ending with a 6,

   (h) numbers ending in a 7 square to a number ending with a 9,

   (i) numbers ending in a 8 square to a number ending with a 4,

   (j) numbers ending in a 9 square to a number ending with a 1,

   So, our previous observation was true.

10. **Pizza for eleven:**

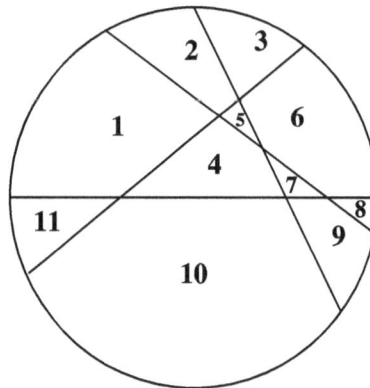

11. **Sum of four consecutive numbers:** Four consecutive numbers will always be two evens and two odds. Since the sum of two evens is even and the sum of two odds is also even, then the sum of any four consecutive numbers is even.

12. **Darts:** There are many possibilities. We start by using as many 16's as possible and then we go down little by little.

    (a) three 16's, sum = 48,

    (b) two 16's and one 4, sum = 36,

    (c) two 16's and one 2, sum = 34,

    (d) two 16's and one 1, sum = 33,

    (e) one 16, and two 4's, sum = 24,

(f) one 16, and two 2's, sum = 20, (g) one 16, and two 1's, sum = 18,
(h) one 16, one 4 and one 2, sum = 22,
(i) one 16, one 4 and one 1, sum = 21,
(j) one 16, one 2 and one 1, sum = 19,
(k) three 4's, sum = 12,
(l) two 4's and one 2, sum = 10,
(m) two 4's and one 1, sum = 9,
(n) one 4 and two 2's, sum = 8,
(o) one 4 and two 1's, sum = 6,
(p) one 4, one 2 and one 1, sum = 7,
(q) three 2's, sum = 6,
(r) two 2's and one 1, sum = 5,
(s) three 1's, sum = 3

13. **Fermat's pet store:** We notice that at the beginning (week zero) there are $128 = 2^7$ cats, in week one there were $64 = 2^6$, in week two he had $32 = 2^5$ cats. If this pattern keeps going then in week three he would have $16 = 2^4$ cats, in week five $8 = 2^3$, in week six $4 = 2^2$, and just 2 in week seven, and just $1 = 2^0$ in week eight... then he would ran out of cats.

14. **Counterfeit coin II:** Two. trying easier cases, for 3 coins only one wights is enough. For 5 coins it will need one or two. If you consider these easier cases you should be able to get a strategy that will help you to need only two weights to find the counterfeit coin out of the nine you started with.

15. **+ and - with pattern blocks:**
Green triangle: ____1/12_____
Red trapezoid: _____1/4_____
Blue parallelogram: ____1/6____
Yellow hexagon: _____1/2_____

Now using them to compute what was asked
(a) $\frac{5}{6} + \frac{1}{2} = 5\ blue + 1\ yellow = 5\ blue + 3\ blue = 8\ blue = \frac{8}{6}$
(b) $1\frac{1}{6} - \frac{3}{4} = 2\ yellow + 1\ blue - 3\ red = 12\ green + 2\ green - 9\ green = 5\ green = \frac{5}{12}$
(c) $1\frac{5}{6} - \frac{3}{2} = 2\ yellow + 5\ blue - 3\ yellow = 5\ blue - 1\ yellow = 5\ blue - 3\ blue = 2\ blue = \frac{2}{6}$

16. **Geoboard I:** (a) 14   (b) 8   (c) 12   (d) 9   (units are in^2, or cm^2, ft^2, etc.)

17. **Geoboard II:** (a) First draw two segments to create a triangle, see first picture below. This triangle has area 6, as it is half of a rectangle with base 4 and height 3. We then break the part of the triangle that is not in the shape we want to find the area of, see second picture below. We get two triangles and one rectangle

The rectangle has area 2 and the rectangles add up to $0.5 + 1 = 1.5$. Thus the area we have to take out is $2 + 1.5 = 3.5$. It follows that the area we do want to count and that is inside that triangle is 2.5.
We now look at the little triangle that is outside the triangle we were working on (look at first picture below). We break it into two smaller triangles that are clearly halves of rectangles, thus the areas are easy to see.

Thus the area is $2.5 + 1 + 0.5 = 4$ (square units).

(b) We first focus on the red rectangle in the first picture below. This rectangle has area 2. Since the piece we are interested in is what is left after taking out two triangles, one with area 0.5 and the other with area 1, then we get that its area is 0.5. We now focus on the rest of the figure and we break it into a triangle and a rectangle (second picture below)

It follows that the area we were looking for is 3 (square units).

c) 19.5   **d)** 21

18. **Sewing:** Reading the problem yields the figure below

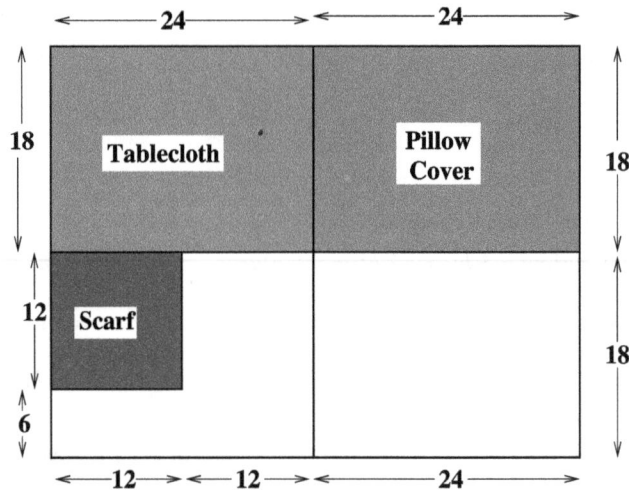

What is left is not colored. So, what is left is half of the cloth minus a 12×12 in^2 piece. Since each quarter of the cloth contains exactly three of these, then the blue part is a sixth of half of the cloth. Since half of the cloth is $6, then the blue part is $1, and thus the non-colored part is $5.

19. **Balancing books:** 3 pounds.

20. **Pattern blocks I:** Pretty much done in class. Could you do 1/5 and 1/7?

21. **Shaded circle:** $\dfrac{11}{15}$.

22. **Pattern blocks II:** We will discuss problems like this in class, and solve many problems that need these types of computations... you should come up with problems for this anyway.

23. **Mind reader I:** $\dfrac{2}{3}$.

24. **Mind reader II:** We set the equation

$$\frac{1}{3}x - \frac{1}{2}x = 8$$

which yields $x = -48$.

25. **Be proper:** The meaning of $2\frac{1}{3}$ is two integers plus a third. Since an integer can be broken intro three thirds, then two integers are six thirds, that plus the other third gives us seven thirds.

26. **A picture is worth a thousand words:** $\frac{1}{2} \cdot \frac{1}{3}$ may be read as a half of a third. Thus we can break the rectangle above into three equal pieces and take one (the one on the left). Then take a a half of that... that is the red square!!

    Can you think of any other interpretations?

27. **Simple math:** $\frac{2}{3}$ of 12 is 8.

    $\frac{2}{3}$ of 14 is $\frac{28}{3}$.... did you do this using pattern blocks? Try it out!

28. **Fighting about pizza:** One of the important things about fractions is to determine what the whole is. So, if we consider the whole to be the total number of slices of pizza (27 pieces) then Frank ate $\frac{12}{27}$ and Dave ate $\frac{15}{27}$. Since

    $$\frac{15}{27} - \frac{12}{27} = \frac{3}{27} = \frac{1}{9}$$

    then Frank ate one-ninth of the total of slices more than Dave.

    Now, if the fractions Frank and Dave are talking about are with a whole that is how much the other ate then Frank ate 3 more slices than Dave, since Dave ate 12 then $\frac{3}{12}$ yields one-fourth. That is what he is talking about! On the other hand, Dave ate 3 less slices than the 15 Frank ate, thus he ate $\frac{3}{15} = \frac{1}{5}$ less than Frank... Can we compare these two facts? They use a different whole for their arguments, so they are both valid **as long as the whole is clearly determined**. In order to decide this argument we need to consider the same whole for both of them. Hence, I would say they are both wrong, and that Frank ate $\frac{1}{9}$ more the pizza than Dave.

29. **Wrong but right:**

    (a) Calculator might help.

    (b) $\frac{1999}{9995}$ $\frac{19999}{99995}$ $\frac{266}{665}$

30. **Grandma's quilt:** Since the total area of the quilt is 30 square feet, and One-third is red, then 10 square feet are red. Similarly, since two-fifths are white then $\frac{2}{5}30 = 12$ square feet are white. Hence, $30 - 10 - 12 = 8$ square feet are blue.

31. **In base 5: (a)** Since $2 + 3$ is 10 in base 5 then we get a zero in the first column on the right. Then we carry one to the second column, which now adds to 6, but this is 11 in base 5 so we keep a 1 in this column and carry a one to the next one. In the third column we get a sum of 5, then we keep a zero and carry a one. Hence,

$$\begin{array}{r} 1\,2\,2_5 \\ +\,3\,3\,3_5 \\ \hline 1\,0\,1\,0_5 \end{array}$$

    (b) In the first column we cannot just subtract we need to borrow one from the column to the left. So, we need to look at $12_5 - 3_5$, which is 4 in base 5. Now the two in the second column becomes a 1, since we want to subtract 3 from 1 then we borrow one from the left, we get $11_5 - 3_5$, which is 3 in base 5. Finally the 2 in the third column becomes a 1 (borrowing!), so the last subtraction is zero. Hence,

$$\begin{array}{r} 2\,2\,2_5 \\ -\,1\,3\,3_5 \\ \hline 0\,3\,4_5 \end{array}$$

    (c) Since 2×3 is 11 in base 5 then the first column has a 1 and we carry a 1 to the next column. In the second column we get another 11 in base 5 plus the 1 carried we get a 2 and carry another one. Thus, in the first multiplication we get 121_5. Since the second multiplication is the same as the previous one, then we get another 121, but shifted. When we add we get

$$
\begin{array}{r}
2\ 2_5 \\
\times\quad 3\ 3_5 \\
\hline
1\ 2\ 1_5 \\
1\ 2\ 1\ 0_5 \\
\hline
1\ 3\ 3\ 1_5 \\
\end{array}
$$

32. **Grade conscious student: Proportions way:** This is direct proportion, as the more you study the better your grades are. The proportion we set is

$$
\frac{3\ days}{5\ points} = \frac{14\ days}{x\ points}
$$

We solve this and get

$$
x = \frac{14 \cdot 5}{3} = \frac{70}{3} \sim 23\ points
$$

Second way: If one studies 3 days to get 5 points then each day of study is going to increase the grade in $\frac{5}{3} = 1\frac{2}{3} = 1.6666666...$ points. Thus after 14 days the grades increase in $14 \cdot 1.6666666...$, which is about 23 points.

33. **Doing the laundry:** 32.

34. **Baking cookies:**
 (a) $2\frac{1}{2}$ cups of flour.
 (b) $\frac{5}{8}$ cups of flour.
 (c) $2\frac{1}{12}$ cups of flour.

35. **Catch the errors:**
 Sam:
 $\frac{3}{4} \times \frac{1}{6} = \frac{9}{12} \times \frac{2}{12} = \frac{18}{12} = \frac{3}{2}$
 Using common denominator for multiplication.
 Tom:
 $\frac{5}{6} \times \frac{3}{8} = \frac{5}{6} \times \frac{8}{3} = \frac{40}{18} = \frac{20}{9}$
 Using 'flipping second fraction' technique from division to perform multiplication.

36. **Chickens:** Since 1.5 chickens lay 1.5 eggs in 1.5 days, then 1.5 chickens lay 1 egg in 1 day. It follows that 1 chicken lays 2/3 of an egg a day.
 (a) Using what we have above we get that 12 chickens lay $12(2/3) = 8$ eggs in a day. So, given 12 days they lay $12 \cdot 8 = 96$ eggs.
 (b) Using what is above we see that 3 chickens would lay $3(2/3) = 2$ eggs in one day. Since we want 24 eggs we need to consider 12 days to get $12 \cdot 2$ eggs. Answer: 12 days.
 (c) Using what is above we see that one chicken would lay $6(2/3) = 4$ eggs in 6 days. Since we want 36 eggs in that amount of time, then we need 9 chickens.

37. **Guests at a dinner:** We set the equations

$$
\frac{x}{2} = R \qquad\qquad \frac{x}{3} = B \qquad\qquad \frac{x}{4} = F
$$

where x is the number of guests, R the number of rice bowls used, B the number of broth bowls used and F the number of fowl bowls used. It follows that

$$\frac{x}{2} + \frac{x}{3} + \frac{x}{4} = 65$$

which has a solution of $x = 60$ guests.

38. **In between:** $\dfrac{2}{5} = \dfrac{140}{350} < \dfrac{141}{350} < \dfrac{142}{350} < \dfrac{143}{350} < \dfrac{150}{350} = \dfrac{3}{7}$

39. **Agnes's running habits:** Note that when the whistle blows Agnes is at $3/8$ of the total length of the bridge. Thus she has $5/8$ of the bridge left. It follows that when she runs toward the far end of the bridge she has to cover $3/8 + 2/8$ of the bridge. Since covering $3/8$ of the bridge (now running towards the train) is exactly when the train reaches the bridge, then when Agnes has $2/8$ of the bridge left to run (towards the far end) then the train is reaching the bridge in the other end.... by assumption they reach the far end of the bridge at the same time. Thus, the train covers the whole bridge in the same time Agnes covers $2/8 = 1/4$ of it. Thus the train goes 4 times faster, at $4 \cdot 10 = 40$ miles per hour.

40. **Birthday cake:** Note that if one cuts the cake in 6 pieces, and Bob takes a fifth of what was left instead of one fourth, then Adnan taking one sixth means that there are five pieces available. Bob takes a fifth, meaning there are four available, Carmen takes a fourth, and thus there are three available, Doreen takes a third and there are two pieces. Hence me and my BFF have exactly the same size piece as everyone else... But Bob took a fourth instead of a fifth. He took more than others!!!

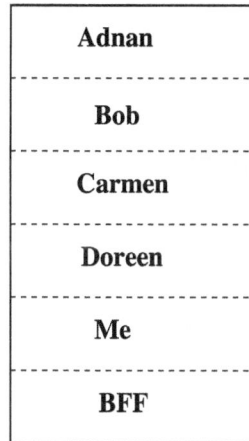

41. **SpongeBob:** 20% of SpongeBob's weight on dry land is $0.2 \cdot 2 = 0.4$ *oz*. This means that his weight with no water is 1.6 *oz*. Now we know that 85% of him is water, thus 15% of his weight is exactly 1.6 *oz*. So, we set the equation

$$\frac{100}{x} = \frac{15}{1.6}$$

we get $x = 160/15 = 32/3 = 10\frac{2}{3}$ *oz*.

42. **Algebra tiles I:** For part (a) the first picture below shows what is given to us, the second shows how to arrange the blocks so they form a rectangle.

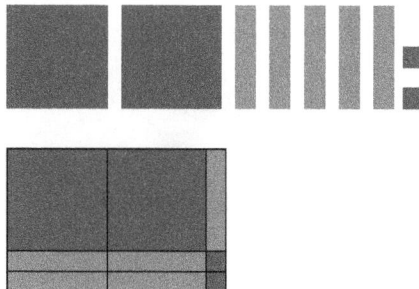

It is easy to see that the rectangle is obtained by multiplying $2x-1$ and $x-2$. It follows that

$$2x^2 - 5x + 2 = (2x-1)(x-2)$$

For part (b) we get

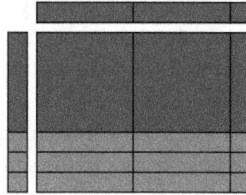

So, $(x-3)(2x+1) = 2x^2 + x - 6x - 3 = 2x^2 - 5x - 3$.

43. **Algebra tiles II:** First we move things around so we get an equation equal to zero. Thus, we need to solve

$$x^2 - 6x + 5 = 0$$

We factor $x^2 - 6x + 5$ in the following figure

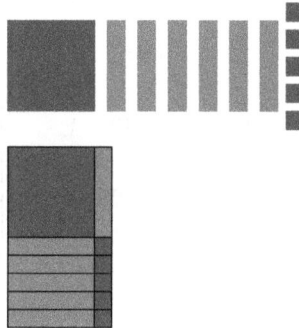

So, we get $x^2 - 6x + 5 = (x-1)(x-5)$. It follows that

$$(x-1)(x-5) = 0$$

and thus $x=1$ or $x=5$.

44. **How old am I?:** Let x be my mom's age and y my age (both current ages). So,

$$x - 15 = 2(y-15)$$

means that 15 years ago my mom was twice as old as I was. we also know that,

$$x = \frac{3}{2}y$$

Plugging the latter equation into the former we get

$$\frac{3}{2}y - 15 = 2(y-15)$$

Multiplying by 2 we get

$$3y - 30 = 4(y-15)$$

which implies $y = 30$.

45. **Isosceles triangle area:** $12\ in^2$
46. **Volume of a trough:** $35\ ft^3$
47. **Length of a ladder:** 7.25 yards
48. **Chevy and Ford:** $45\ mi/hr$
49. **Width of a river:** $30\sqrt{3}$ yards
50. **Irrigation problem:** $(100 - 25\pi)\%$
51. **Zooming in:** 24 cm by 36 cm
52. **Squares in squares in squares:** $\dfrac{1}{32}$
53. **Star-shaped area:** $4 - \pi$

References

1. Burger, W. and Shaughnessy, J., *Characterizing the van Hiele Levels of Development in Geometry.* Journal for Research in Mathematics Education 17: pp. 31–48. 1986
2. Ma, L. *Knowing and Teaching Elementary Mathematics: Teachers' Understanding of Fundamental Mathematics in China and the United States.* Mahwah, NJ: Lawrence Erlbaum Associates, 1999.
3. McAnelly, N., BEYOND THE GEOMETRY: DISCOVERING HOW GEOMETRIC THINKING DEVELOPS, NCTM, Indianapolis, IN. 2011.
4. Polya, G. *How to Solve It.* Princeton, NJ: Princeton University Press, 1945.
5. Van de Walle, J., Karp, K., & Bay-Williams, J. *Elementary and Middle School Mathematics: Teaching Developmentally* (7th ed.) Boston, MA: Pearson Education Inc., 2008.

www.ingramcontent.com/pod-product-compliance
Lightning Source LLC
Chambersburg PA
CBHW082035230326

41598CB00081B/6520